Global Material Resources Outlook to 2060

ECONOMIC DRIVERS AND ENVIRONMENTAL CONSEQUENCES

This work is published under the responsibility of the Secretary-General of the OECD. The opinions expressed and arguments employed herein do not necessarily reflect the official views of OECD member countries.

This document, as well as any data and any map included herein, are without prejudice to the status of or sovereignty over any territory, to the delimitation of international frontiers and boundaries and to the name of any territory, city or area.

Please cite this publication as:
OECD (2019), *Global Material Resources Outlook to 2060: Economic Drivers and Environmental Consequences*, OECD Publishing, Paris.
https://doi.org/10.1787/9789264307452-en

ISBN 978-92-64-30744-5 (print)
ISBN 978-92-64-30745-2 (pdf)
ISBN 978-92-64-32828-0 (HTML)
ISBN 978-92-64-39828-3 (epub)

The statistical data for Israel are supplied by and under the responsibility of the relevant Israeli authorities. The use of such data by the OECD is without prejudice to the status of the Golan Heights, East Jerusalem and Israeli settlements in the West Bank under the terms of international law.

Photo credits: Cover © DR Travel Photo and Video/Shutterstock.com

Corrigenda to OECD publications may be found on line at: *www.oecd.org/about/publishing/corrigenda.htm*.
© OECD 2019

You can copy, download or print OECD content for your own use, and you can include excerpts from OECD publications, databases and multimedia products in your own documents, presentations, blogs, websites and teaching materials, provided that suitable acknowledgement of OECD as source and copyright owner is given. All requests for public or commercial use and translation rights should be submitted to *rights@oecd.org*. Requests for permission to photocopy portions of this material for public or commercial use shall be addressed directly to the Copyright Clearance Center (CCC) at *info@copyright.com* or the Centre français d'exploitation du droit de copie (CFC) at *contact@cfcopies.com*.

Foreword

As nations become wealthier, does this necessarily lead to a proportional increase in the weight of the materials they consume? Can economies successfully decouple materials use from economic growth? These pressing questions lie at the heart of national and international discussions about the transition to a more resource efficient, circular economy.

Globally, the use of material resources grew from 27 billion tonnes (Gt) in 1970 to 89 billion tonnes in 2017. This will increase further with continued population growth and economic development. However, such growth in materials use, coupled with the environmental consequences of material extraction, processing and waste, is likely to increase the pressure on the resource bases of our economies and jeopardise future gains in well-being.

Many OECD countries and emerging economies are implementing policies to stimulate the transition to circular economies. Resource efficiency is also central to our efforts to deliver on the Paris Agreement and achieve the Sustainable Development Goals, especially those related to climate change, biodiversity, water, energy and responsible consumption and production. In addition, global fora like the G7 and the G20 are increasingly prioritising resource efficiency in their discussions.

The *Global Material Resources Outlook to 2060* seeks to support these efforts through projections on the future use of material resources. It is the first report that provides an in-depth examination of the likely path of future materials use and the economic drivers that underpin these projections. The report focuses on three major socioeconomic trends that influence the use of material resources: the expected growth in emerging and developing economies relying on material-intensive infrastructure development; the evolution of many economies towards a greater role for the services sector; and the accelerating transformation of production technologies and processes. These trends will have differing, and sometimes conflicting, impacts on the scale and intensity of future materials use.

This *Outlook* projects that, in the absence of new policies, global materials use would rise from 89 Gt in 2017 to 167 Gt in 2060. This growth is reflected in all major categories of materials: metallic ores (9 to 20 Gt), non-metallic minerals (44 to 86 Gt), biomass (22 to 37 Gt) and fossil fuels (15 to 24 Gt). In addition, the extraction, processing and disposal of materials brings significant environmental consequences, which will be magnified as materials use increases. These include a doubling of greenhouse gas emissions, pollution to the soil, water and air, and toxic effects on humans and aquatic and terrestrial ecosystems. Such environmental consequences will hurt our economies and our societies.

This *Outlook* can help decision-makers understand the direction in which we are heading and help to assess which policies can support a more circular economy. The OECD stands ready to assist governments in making this transition by designing, developing and delivering better resource efficiency policies for better lives.

Angel Gurría
OECD Secretary-General

Acknowledgements

This report was managed by Shardul Agrawala, Head of the Environment and Economy Integration Division of the OECD Environment Directorate. Ruben Bibas, Jean Chateau, Rob Dellink and Elisa Lanzi of the OECD Environment Directorate prepared the report and carried out the underlying modelling analysis. Martin Benkovic provided modelling support and created the graphics in the report. Peter Börkey provided guidance on the policy aspects of resource efficiency and the transition to a circular economy.

Stefan Giljum and Mirko Lieber (Vienna University, Austria) provided the material flow data. Jinxue Hu and Ton Bastein (TNO Research, The Netherlands) provided data on recycling and reprocessing. Frank Pothen (Leibniz Universität Hannover, Germany) provided information on materials decoupling. Lisa Congyuan Yao and Silpa Kaza (World Bank) provided the analysis on municipal waste generation. Alvaro Calzadilla, Victor Nechifor and Matthew Winning (University College London, United Kingdom) provided the case study analysis on iron and steel. Matthias Pfaff and Marcel Soulier (Fraunhofer ISI, Germany) provided the case study analysis on copper. Lauran van Oers and Ester van der Voet (Leiden University, The Netherlands) provided the environmental impact assessment on metals and construction materials.

Katjusha Boffa, Aziza Perrière and Jack McNeill (all OECD Environment Directorate) provided administrative support. Beth del Bourgo, Jessica Li and Tobias Udsholt (all OECD Environment Directorate) created communication materials for the report. The report was edited by Fiona Hinchcliffe with guidance from Janine Treves (OECD Public Affairs and Communications Directorate).

The OECD Environment Policy Committee (EPOC) was responsible for the oversight of the development of the report. In addition, the Working Party on Resource Productivity and Waste (WPRPW) and the Working Party on Integrating Environmental and Economic Policies (WPIEEP) and the experts that participated in the RE-CIRCLE Technical Workshops reviewed earlier drafts.

The report also benefited from expert input and feedback on earlier drafts by Elisabetta Cornago, Anthony Cox, Frithjof Laubinger, Myriam Linster, and Andrew McCarthy (all OECD Environment Directorate), Araceli Fernandez Pales (International Energy Agency) as well as Raimund Bleischwitz (University College London, United Kingdom), Tara Caetano (University of Cape Town, South Africa), Martin Distelkamp (GWS, Germany), Stefan Giljum (Vienna University of Economics and Business, Austria) and Steve Hatfield-Dodds (ABARES, Australia).

Finally, this work would not have been possible without the financial support from France, Luxembourg, the Netherlands, Switzerland and the United Kingdom.

Table of contents

Foreword 3
Acknowledgements 5
Abbreviations and acronyms 13
Executive summary 15
    What are the key projected trends for materials use? 15
    What are the environmental consequences? 16
    What are the policy implications? 16

Chapter 1. **An overview of global material resource use to 2060** 17
    1.1. Global economic growth relies on an increased use of material resources 18
    1.2. Materials use is set to keep growing without new policies 19
    1.3. How is uncertainty accounted for? 22
    1.4. The projections imply large environmental and sustainable development challenges 23
    1.5. Improving materials efficiency will require better policies 26
    1.6. How is the report structured? 27
    Notes 28
    References 29

Part I. The economic drivers of materials use 31

Chapter 2. **Modelling future materials use and its economic drivers** 33
    2.1. Introduction 34
    2.2. How is future materials use modelled? 37
    2.3. What is the baseline scenario approach? 48
    Notes 50
    References 52
    Annex 2.A. The modelling framework 55
    2.A.1 The modelling approach to produce economic and materials projections 55
    2.A.2 The ENV-Growth model 55
    2.A.3 The ENV-Linkages model 57

Chapter 3. **Projecting the economic baseline scenario** 61
    3.1. Population and economic growth will see major regional shifts 64
    3.2. The services sector will drive demand growth 67
    3.3. Production processes will rely more on new technologies and services 70
    3.4. Several areas of uncertainty affect the socioeconomic projections 73
    Notes 77
    References 78
    Annex 3.A. Detailed results and supplementary materials 79
    3.A.1 Detailed total population projections and the ageing process 79

3.A.2 Detailed GDP growth and assumptions about drivers of the projected GDP per capita 82

Chapter 4. **Projections of the economic drivers of materials use** 87

4.1. Economic development and construction materials are closely linked 90
4.2. The global rise of services helps reduce materials use 96
4.3. There is a gradual transition away from primary material inputs 99
4.4. Demand for materials declines as economies mature 105
Notes 109
References 110
Annex 4.A. Detailed results and supplementary materials 112

Part II. Materials use to 2060 115

Chapter 5. **Projections of primary materials use** 117

5.1. Development levels affect materials extraction rates 120
5.2. Materials extraction is projected to almost double by 2060 121
5.3. Materials intensity is projected to decline 128
5.4. Materials use per capita is projected to increase in most countries 131
5.5. Socioeconomic scenarios are a source of uncertainty 135
Notes 138
References 138
Annex 5.A. Detailed results and supplementary materials for Chapter 5 139

Chapter 6. **Projections of recycling and secondary materials** 141

6.1. Secondary materials are only a modest part of total materials use 144
6.2. Recycling is projected to triple 148
6.3. Secondary metal production is projected to grow as fast as primary metal production 153
6.4. Uncertainty surrounds the recycling and secondary materials projections 155
Notes 156
References 156
Annex 6.A. Detailed results and supplementary materials 158

Chapter 7. **Case studies on demand and supply risks for specific materials: copper, iron and steel and critical materials** 161

7.1. Detailed case studies bring global patterns into sharper focus 164
7.2. The copper case highlights the close links between primary and secondary material flows 164
7.3. The case of iron and steel reveals very large recycling potential 168
7.4. The case of critical materials in the OECD shows the key role of supply risks 173
Notes 177
References 178

Part III. The environmental consequences of materials use 179

Chapter 8. **Projections of the environmental consequences of materials use** 181

8.1. Materials use has many environmental consequences 184
8.2. Reducing greenhouse gas emissions is strongly linked to materials use policies 186
8.3. The increase in materials use will exacerbate environmental impacts 188
Notes 199
References 200
Annex 8.A. Detailed results and supplementary materials 203
8.A.1. Comparing the GHG projections to the literature 203

8.A.2. Climate damages 204
8.A.3. Detailed explanations of the analysis of environmental impacts 205
8.A.4. Detailed results for projected environmental impacts 208

Tables

Table 2.1. ENV-Linkages model regions 39
Table 2.2. Material flows linked to sectoral output 41
Table 2.3. Material flows linked to inputs into processing sector 42
Table 3.1. Input composition for the production of manufacturing goods 73
Table 3.2. Uncertainty surrounding the main macroeconomic variables in 2060 76
Table 4.1. Projected input composition for the production of manufacturing goods 99
Table 4.2. Structural changes affect the demand structure for metals 101
Table 5.1. Projections of global materials extraction in the central baseline scenario 124
Table 5.2. Projected growth of global materials use in the central baseline scenario is lower than in the 2017 UNEP-IRP report 125
Table 5.3. Materials intensity improvements compared with historical rates and other studies .. 129
Table 6.1. Estimates of current recycling rates and recycled content of metals 145
Table 7.1. Criticality of selected materials in the OECD in 2012 174
Table 8.1. Potential environmental impacts by material group 184
Table 8.2. Environmental impact categories and indicators 189

Table 2.A.1. Sectoral aggregation of ENV-Linkages 59
Table 3.A.1. Population by region: historical and projected trends 79
Table 3.A.2. Real GDP by region: historical and projected trends 82
Table 8.A.1. Projected environmental impacts of selected metals in 2060 208

Figures

Figure 1.1. GDP grows faster in developing countries than in the OECD and the BRIICS by 2030 18
Figure 1.2. Structural and technology change is projected to slow down the growth in materials use . 20
Figure 1.3. GDP is projected to grow more quickly than materials use and recycling more quickly than mining 21
Figure 1.4. Growth in materials use is projected for all regions 22
Figure 1.5. Uncertainties on materials use is especially large for non-metallic minerals 23
Figure 1.6. Global environmental impacts differ significantly across materials 24
Figure 1.7. Roadmap of the report 28
Figure 2.1. The scenario projections build on a complex modelling framework 38
Figure 2.2. Splitting the recycling and processing sectors provides relevant detail 43
Figure 2.3. Historic estimates of key indicators indicate no absolute supply scarcity for selected metals 47
Figure 2.4. Fossil fuel use and CO2 emissions are projected to increase in the IEA CPS scenario 49
Figure 3.1. GDP is projected to increase significantly in all countries but global growth is slowing down 62
Figure 3.2. World population is projected to keep growing but less rapidly than in the past 64
Figure 3.3. Living standards are projected to gradually converge 65
Figure 3.4. Emerging economies drive the projected global GDP growth 67
Figure 3.5. Demand for services is projected to increase more than the economy-wide average 69
Figure 3.6. The share of household consumption in GDP is projected to increase with ageing 71

Figure 3.7. Labour efficiency and capital supply drive per-capita GDP growth 72
Figure 3.8. Population and income convergence assumptions lead to long-term uncertainties in GDP 74
Figure 3.9. Uncertainties surrounding population and income convergence assumptions are greatest for emerging economies 75
Figure 4.1. Global growth is faster in less materials-intense sectors 88
Figure 4.2. Emerging Asian and Sub-Saharan African economies are projected to replace China as engines of global growth 91
Figure 4.3. In most countries investment is projected to increase over time 93
Figure 4.4. Construction activity is linked to investment booms 94
Figure 4.5. Non-OECD Asia and Africa are projected to see the strongest growth of construction materials use 95
Figure 4.6. Projected shifts in regional demand differs from the projected shifts in regional production 97
Figure 4.7. Less materials-intense services are projected to see above-average growth 98
Figure 4.8. Energy efficiency, renewables and electrification partially decouple fossil fuel use from GDP 100
Figure 4.9. Metals production is only projected to decouple from mining inputs after 2030 103
Figure 4.10. The costs of recycling are projected to fall compared to the costs of mining 104
Figure 4.11. Secondary non-ferrous metal processing tends to be more labour-intensive than primary 105
Figure 4.12. GDP and floor space per capita have moved in the same direction between 2000 and 2016 107
Figure 4.13. Construction materials use in China is more affected by sectoral assumptions than by the socioeconomic uncertainties 108
Figure 5.1. Materials intensity is projected to decrease by 2060 119
Figure 5.2. Non-metallic minerals constitute the bulk of materials extraction 120
Figure 5.3. Materials use is heterogeneous across regions and development levels 121
Figure 5.4. Global materials extraction is projected to increase across all material types 122
Figure 5.5. Construction materials use is projected to almost double between 2017 and 2060, with the largest growth in sand, gravel and crushed rock 123
Figure 5.6. Global use of each material is projected to grow at a specific rate 126
Figure 5.7. Materials use is projected to increase in most countries 127
Figure 5.8. Global materials intensity is projected to decrease by 1.3% per year on average, and especially strongly in emerging economies 128
Figure 5.9. Materials intensity is projected to decrease everywhere, but not at the same pace 130
Figure 5.10. Global materials use per capita is projected to keep increasing over time 132
Figure 5.11. Materials use per capita is projected to increase in most countries 133
Figure 5.12. Construction materials use per capita is projected to increase in most countries 134
Figure 5.13. Uncertainties on population and income convergence strongly impact materials use projections 136
Figure 5.14. Materials intensity projections are less impacted by population and income convergence uncertainties 137
Figure 6.1. The recycling sector is projected to outpace the mining sector 143
Figure 6.2. The share of secondary metals is very heterogeneous across selected metals 146
Figure 6.3. Many sectors use recycling as input 147
Figure 6.4. The recycling sector is projected to outpace the mining sector 148
Figure 6.5. In almost all regions, recycling is projected to grow more rapidly than mining 149
Figure 6.6. The recycling sector is projected to remain small in all regions 150
Figure 6.7. Projected municipal waste generation increases in all regions to 2050 152

Figure 6.8. The share of secondary metal production is projected to remain roughly unchanged until 2060 153
Figure 6.9. The relative price of secondary non-ferrous metals is projected to increase 154
Figure 6.10. Mining and recycling output vary with population and income convergence assumptions in proportion to GDP 155
Figure 6.11. The share of secondary metals is projected to change little under alternative population and income convergence assumptions... 156
Figure 7.1. Copper material flows in 2015 differed significantly across countries............................. 165
Figure 7.2. Copper stocks influence waste generation and recycling volumes 167
Figure 7.3. Per-capita stocks of steel have stabilised in some developed countries............................ 169
Figure 7.4. Global steel production is projected to grow significantly 169
Figure 7.5. Global primary and secondary steel production are both projected to grow significantly 170
Figure 7.6. Growth in per-capita steel stocks is projected to continue in China and India 172
Figure 7.7. Steel scrap production is projected to grow faster than iron ore production..................... 173
Figure 7.8. If production shifts towards countries with large reserves, only few materials remain critical in the OECD by 2030 176
Figure 8.1. Emissions from materials management are projected to increase, but not more rapidly than other emission sources.. 183
Figure 8.2. Global GHG emissions are projected to grow in the central baseline scenario 186
Figure 8.3. Greenhouse gas emissions are projected to increase... 187
Figure 8.4. Per kg environmental impacts are higher for primary than for secondary materials 190
Figure 8.5. Secondary materials lead to much lower environmental impacts 192
Figure 8.6. Per kg environmental impacts decrease over time, except when ore grades degrade significantly... 193
Figure 8.7. Environmental impacts of selected metals are in most cases projected to more than double by 2060 195
Figure 8.8. Projected environmental impacts of selected materials in 2060 197
Figure 8.9. Metals and concrete represent a significant share of the environmental impacts 198

Figure 2.A.1. Stylised representation of conditional income convergence 56
Figure 2.A.2. Global distribution of per-capita income 56
Figure 2.A.3. The ENV-Growth Modelling framework overview.. 57
Figure 3.A.1. Shares of children and elderly in total population.. 80
Figure 3.A.2. Growth of the working age population 81
Figure 3.A.3. GDP per capita evolution between 2017 and 2060 83
Figure 3.A.4. Changes in the drivers of employment rates 84
Figure 3.A.5. Evolution of net-investment financial components.. 85
Figure 3.A.6. Evolution of capital to GDP ratios 86
Figure 4.A.1. The composition of investment expenditure by commodity 112
Figure 4.A.2. The production of the construction sector by region.. 113
Figure 4.A.3. Construction materials intensity of GDP by region 114
Figure 5.A.1. Projections of materials use by sector... 139
Figure 6.A.1. Global production and demand structure of the recycling sector................................. 159
Figure 6.A.2. Production tax rates for recycling 160
Figure 8.A.1. Global GHG emissions projections compared to the literature.................................... 203
Figure 8.A.2. Damages from climate change 204
Figure 8.A.3. Environmental impacts of selected metals are in most cases projected to more than double by 2060 209

Boxes

Box 1.1. OECD Policy Guidance on Resource Efficiency 27
Box 2.1. Key definitions in the report 35
Box 2.2. Policies for resource efficiency and a transition to a circular economy 37
Box 2.3. Understanding decoupling 45
Box 2.4. The IEA's World Energy Outlook scenarios 49
Box 3.1. The macroeconomic impacts of ageing 70
Box 4.1. Energy efficiency and fossil fuel use 100
Box 5.1. Global extraction projections in the context of the literature 124
Box 5.2. Materials intensity projections in the context of the literature 129
Box 6.1. Representing informal sectors is challenging 147
Box 6.2. The potential for increasing recycling rates 151
Box 6.3. Municipal solid waste is a growing issue 152
Box 8.1. The rise and rise of plastics 185

Abbreviations and acronyms

BF-BOF	Blast-Furnace/Basic Oxygen Furnace
BRIICS	Brazil, Russia, India, Indonesia, China and South Africa
CES	Constant Elasticity of Substitution
CFC	Chlorofluorocarbons
CGE	Computable General Equilibrium
CO_2	Carbon Dioxide
CO_2-eq	Carbon Dioxide equivalent
CPS	Current Policies Scenario (IEA World Energy Outlook)
DMC	Domestic Material Consumption
EAF	Electric Arc Furnace
EoL	End of Life
EPR	Extended Producer Responsibility
EU	European Union
GDP	Gross Domestic Product
GHG	Greenhouse Gas
Gt	Gigatonnes (billion metric tonnes)
GTAP	Global Trade Analysis Project
GVC	Global Value Chain
HFC	Hydrofluorocarbons
ICT	Information and Communication Technologies
IEA	International Energy Agency
IMF	International Monetary Fund
IPCC	Intergovernmental Panel on Climate Change
IRP	International Resource Panel (UNEP)
LCA	Life Cycle Analysis
LULUCF	Land Use, Land Use Change and Forestry
MFA	Material Flow Analysis
MSW	Municipal Solid Waste

ABBREVIATIONS AND ACRONYMS

NDC	Nationally Determined Contributions
n.e.c.	Not elsewhere classified
NO_x	Nitrous Oxides
NPS	New Policies Scenario (IEA World Energy Outlook)
PFC	Perfluorocarbons
PPP	Purchasing Power Parities
RMC	Raw Material Consumption
R&D	Research and Development
SAM	Social Accounting Matrix
SDS	Sustainable Development Scenario (IEA World Energy Outlook)
SF_6	Sulphur Hexafluoride
SO_2	Sulphur Dioxide
STAN	OECD's Structural Analysis Database
TFP	Total Factor Productivity
UN	United Nations
UNEP	United Nations Environment Programme
USD	United States Dollars
WU	Vienna University of Economics and Business
WIOD	World Input-Output Database

Executive summary

In the coming decades, growing populations with higher incomes will drive a strong increase in global demand for goods and services, and, as a result, for the material resources to support this growth. Although global population growth is projected to slow down, global population is projected to rise to more than 10 billion by 2060. Over the same period, living standards are gradually converging across economies. Thus, emerging and developing economies will grow faster than countries in the OECD region.

- The OECD ENV-Linkages model projects that global GDP will more than triple between 2017 and 2060 in the central baseline scenario. The global average per capita income is projected to reach USD 37 000 by 2060, almost as high as the current OECD level.
- Production and consumption are shifting towards emerging and developing economies, which on average have higher – but declining – materials intensity.
- The growing share of services in the economy will slow down the growth in materials use, as the materials intensity of services is lower than that of agriculture or industry.
- Technological developments will help decouple growth in production levels from the material inputs to production.

What are the key projected trends for materials use?

This report presents an outlook for global materials use to 2060. It explores how socioeconomic trends drive changes in the use of different materials. The report also delves into the various environmental consequences from the production and use of materials. It provides global, sectoral and regional trends for the use of 60 different materials (including metals, non-metallic minerals, fossil fuels and biomass), assuming that today's policies remain unchanged. The report presents projections for both primary materials and secondary materials.

- Global primary materials use is projected to almost double from 89 gigatonnes (Gt) in 2017 to 167 Gt in 2060. Non-metallic minerals – such as sand, gravel and limestone – represent the largest share of total materials use. These non-metallic minerals are projected to grow from 44 Gt to 86 Gt between 2017 and 2060. Metal use is smaller when measured in weight, but is projected to grow more rapidly and metal extraction and processing is associated with large environmental impacts.
- The strongest growth in materials use is projected to occur in emerging and developing economies. China remains the largest consumer, but the central baseline scenario projects a rapid stabilisation of steel and construction materials use in China. Other non-OECD countries – such as India, Indonesia, and most countries in Sub-Saharan Africa and Asia – are projected to undergo an economic and materials use growth spurt. Even in the OECD, where economic growth rates are more modest, materials use grows between 1% and 2% per year on average.

- The materials intensity of the global economy is projected to decline more rapidly than in recent decades – at a rate of 1.3% per year on average. This stems from the following trends: the global economy orients towards more services, technologies become more efficient, and the construction boom in China phases out.
- This decline in material intensity reflects a relative decoupling: global materials use increases, but not as fast as GDP.
- Recycling is projected to gradually become more competitive compared to extraction of primary materials, leading the recycling sector to outpace growth in mining.
- The strong increase in demand for materials implies that both primary and secondary materials use increase at roughly the same speed. The relatively high labour costs for secondary production technologies hampers further penetration of secondary materials, despite the competitiveness increases in recycling.

What are the environmental consequences?

- Global GHG emissions from all sources are projected to reach 50 Gt CO_2-eq. before 2030, and rise to 75 Gt CO_2-eq. by 2060. The bulk of total emissions remains CO_2 emissions from fossil fuel combustion. The ambitions of the Paris Climate Agreement, including the Nationally Determined Contributions (NDCs) and the "well below two degrees" objective, are thus not met under the central baseline scenario. Materials management activities are responsible for two thirds of GHG emissions. Greenhouse gas (GHG) emissions related to materials management will rise from 30 Gt CO_2-equivalents (CO_2-eq.) in 2017 to about 50 Gt CO_2-eq. by 2060.
- Fossil fuel use and the production and use of iron and steel and construction materials lead to large energy-related emissions of greenhouse gases and air pollutants. The volume of concrete use is so large that even relatively low per-kg impacts imply large consequences: concrete production account for 12% of total GHG emissions in 2060, and the production of metals for 12%.
- Metals extraction and use have a wide range of environmental consequences, including toxic effects on humans and ecosystems. The overall environmental impacts of extraction and processing of key metals are projected to at least double between 2017 and 2060, mostly driven by the increase in the scale of materials use.
- The per-kg environmental impacts of secondary materials are estimated to be an order of magnitude lower than those of primary materials. Policies that further ramp up the transition to secondary materials use and promote circularity will thus lead to overall reductions in environmental impacts.

What are the policy implications?

Improving resource efficiency and stimulating the transition to a circular economy is key to address the wide range of environmental consequences linked to materials use, as well as policy objectives related to security of resource supply and creating jobs. Governments face the complex challenge of designing policy packages to that end, while ensuring coherence with other policy domains such as trade and innovation policies. Such a policy package could also contribute to achieving the Sustainable Development Goals.

Chapter 1.

An overview of global material resource use to 2060

This chapter presents the main insights from the report and puts them into the wider policy context. The first section links the projections of the economic drivers to the projected growth in materials use between 2017 and 2060. The second section discusses the environmental consequences of these materials use projections and links them to the Sustainable Development Goals. The final section presents policy insights drawn from the analysis.

1.1. Global economic growth relies on an increased use of material resources

The world has seen strong economic developments in recent decades. Global economic growth has been underpinned by strong increases in the use of material resources. Materials are used in almost all parts of the economy, not least in construction, the energy sector, and manufacturing. Growth in these sectors continues to rely on increasing the material resource inputs. A detailed understanding of the evolution of the global economy and how different economic activities link to the use of different materials is essential for an understanding of future materials use.

Global population growth is projected to slow down, but nonetheless, the 2017 total of 7.5 billion is projected to grow with another 2.7 billion people by 2060 (UN, 2017[1]). At the same time, living standards will continue to advance in all countries, and – conditional on national circumstances – gradually convergence toward those of the most advanced countries (hereafter labelled income convergence): the growth rates of Gross Domestic Product (GDP) per capita tend to be higher in emerging and developing economies than in the OECD region. Between 2017 and 2060, the global average GDP per capita is projected to triple and to reach the current level of the OECD.

Population growth and income convergence together drive the growth of the global economy. The projected increase in population and global per capita income levels will result in a more than tripling of global GDP. Large populations and rapid catching up of living standards in the People's Republic of China (hereafter China), and to a lesser extent in India and the rest of Southeast Asia, will drive global growth the most in the coming decades.

However, global growth is projected to be lower than in the past. The annual global GDP growth rate is projected to stabilise below 2.5% per year (Figure 1.1), a full percent-point below the average at the turn of the 21st century.[1]

Figure 1.1. GDP grows faster in developing countries than in the OECD and the BRIICS by 2030

Annual growth rates of GDP (2011 PPP)

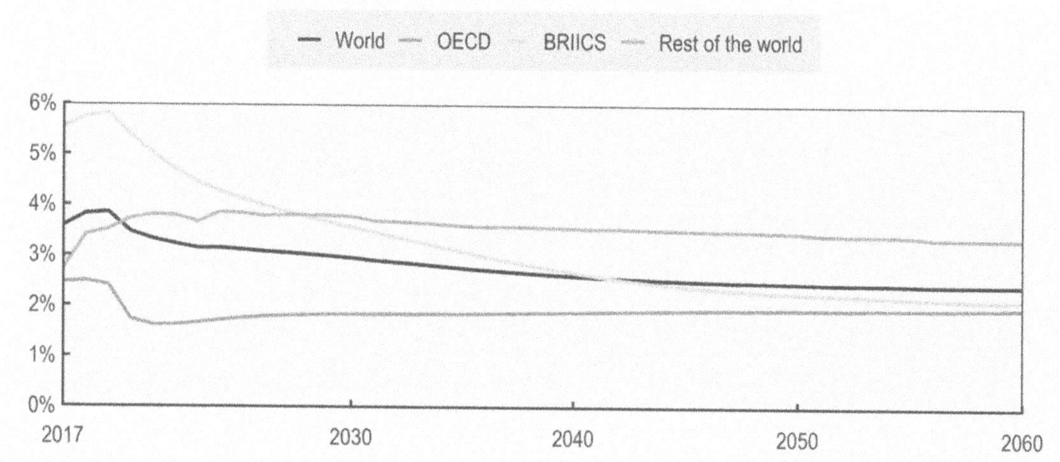

Source: ENV-Growth model (OECD Environment Directorate) and OECD Economics Department (Guillemette and Turner, 2018[2]).

StatLink https://doi.org/10.1787/888933884270

A key driver of this slowdown is the decline in the growth rate of China, which is only partially offset by strong growth in other emerging economies such as India, followed by high growth in large parts of Sub-Saharan Africa. Growth in the OECD countries is projected to remain fairly stable at a little less than 2% on average between 2017 and 2060.

Strong links between economic growth, investment, infrastructure and construction drive an increase in global materials use. As the economies of fast-growing countries mature and develop infrastructure, their use of non-metallic minerals and metals increases strongly. This has been occurring in China in the past two decades. While in China the demand for construction materials will stabilise as the construction boom comes to an end, it will increase in many Asian and African countries in the coming decades.

Economies also continue to undergo structural change, with consequent changes in the contribution of the different sectors to the economy. This affects all countries regardless of their development phase. Trends such as income growth, digitalisation and ageing imply an increasing share of services sectors in the economy. This increased share of services holds not only for final demand (demand for services by households and government), but also for the input of services in industry. The global share of services in total economic activity is projected to increase from 50% to 53.5% between 2017 and 2060. The increase is larger in non-OECD countries (from 37% to 44%) than in the OECD region (from 59% to 64%). As the services sectors have lower materials intensity (materials use per unit of output) than agriculture and industry, this shift towards services can improve the materials productivity of the economy by 2060.[2]

Technology improvements limit the growth in future materials use, despite production growth. These reductions in materials intensity are projected to occur in all major sectors of the economy, albeit at varying rates.

Technological developments can also reduce the cost of recycling relative to the costs of mining raw materials. However, the share of secondary materials in the economy is not projected to change significantly because of the high labour costs involved. Production processes that rely on secondary materials tend to be more labour-intensive than those using primary materials – at least for non-ferrous metals – thus high labour costs are projected to hold back the increase in the share of secondary materials production.

1.2. Materials use is set to keep growing without new policies

In the baseline scenario, the use of primary materials is projected to roughly double from 89 Gt in 2017 to 167 Gt in 2060. The use of all materials categories considered in the analysis (biomass, fossil fuels, metals and non-metallic minerals) will increase. The projected growth boost for emerging and developing economies in particular drives a materials-intensive boost in infrastructure (and construction).

At the global level, structural and technology changes can mitigate the increase in materials use driven by economic growth. The effects of income convergence, structural and technology changes are presented for the central baseline scenario in Figure 1.2. First, economic growth and convergence in income levels across countries can strongly influence global material resources projections, especially the share of materials use by emerging and developing countries in the global economy. This leads to higher growth in global materials use. Indeed, with fixed materials intensity at the regional level and the baseline economic growth projections, all else equal, materials use would grow to more than 300 Gt by 2060 (see the second bar in Figure 1.2).

Structural change, in particular the shift towards services projected for all regions, can put brakes on the strong growth in materials use – reducing materials use by 80 Gt by 2060 (third bar in the figure). Technology developments within sectors, for instance the use of more efficient technologies in production processes, would save another 68 Gt of materials from being used (fourth bar).

Taken together, these economic drivers are projected to increase global materials use while significantly reducing materials intensity. Thus, the central baseline scenario, in which all economic drivers are combined, projects an increase in total materials use between 2017 and 2060 of 78 Gt (last bar in Figure 1.2), i.e. less than doubling. In contrast, if existing economic activity was simply scaled up, materials use could more triple.

Figure 1.2. Structural and technology change is projected to slow down the growth in materials use

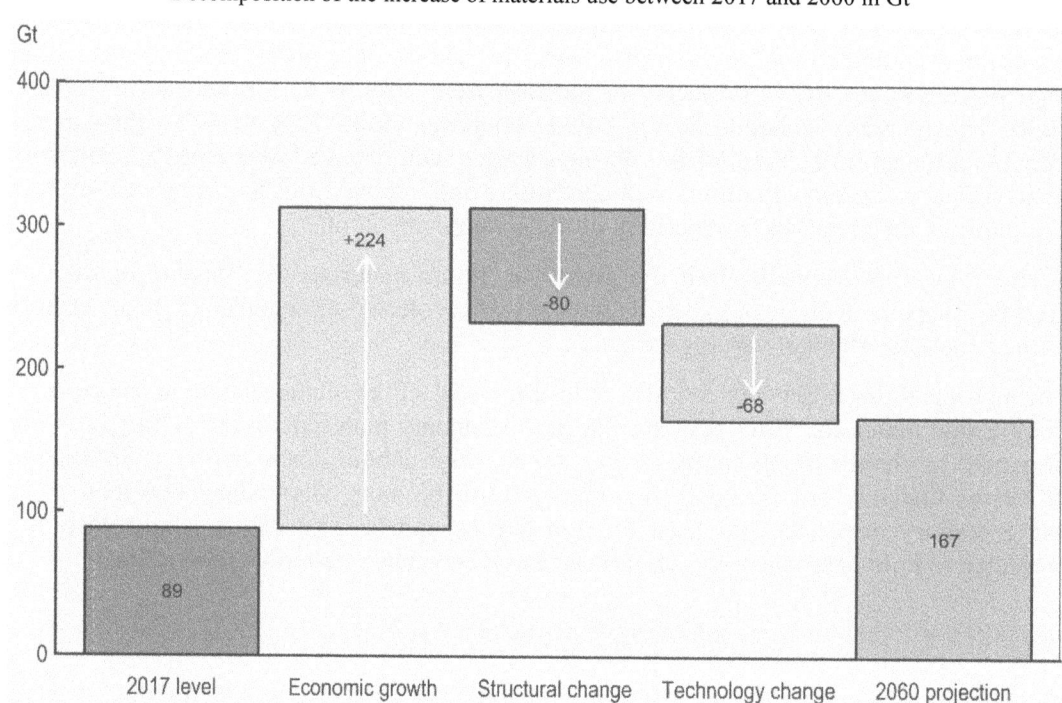

Decomposition of the increase of materials use between 2017 and 2060 in Gt

Note: The four bars read as follows (from left to right):
1. *Economic growth* represents a counterfactual projection in which materials use is assumed to grow at the same speed as GDP and thus in which the regional materials intensity of GDP stays constant.
2. *Structural change* identifies the contribution of sectoral shifts to reducing global materials use by differentiating sectoral growth rates.
3. *Technology change* identifies the contribution of technology improvements to reducing global materials use by differentiating growth rates of materials inputs to sectoral output.
4. The combined effects lead to the *Central baseline projected growth*.

Source: OECD ENV-Linkages model.

StatLink https://doi.org/10.1787/888933884289

Global GDP is projected to more than triple over the period considered, while materials use less than doubles. Figure 1.3 (Panel A) shows the difference in projected growth of GDP and materials use in the central baseline scenario. The gap between growth of GDP and materials use reflects total (economic) decoupling, i.e. the combined effect of structural change and production technology changes shown in Figure 1.2.

The increase in the use of secondary materials also contributes to achieving decoupling of GDP growth and materials use. Panel B of Figure 1.3 highlights that the output growth of the mining sector is constrained by the ongoing decoupling and is thus projected to grow relatively slowly. In contrast, the reprocessing (e.g creating steel from scrap) and especially recycling of materials (which includes recycling of glass, paper and other processed materials) are projected to grow more rapidly.

Figure 1.3. GDP is projected to grow more quickly than materials use and recycling more quickly than mining

Evolution of selected global variables, index 1 in 2017

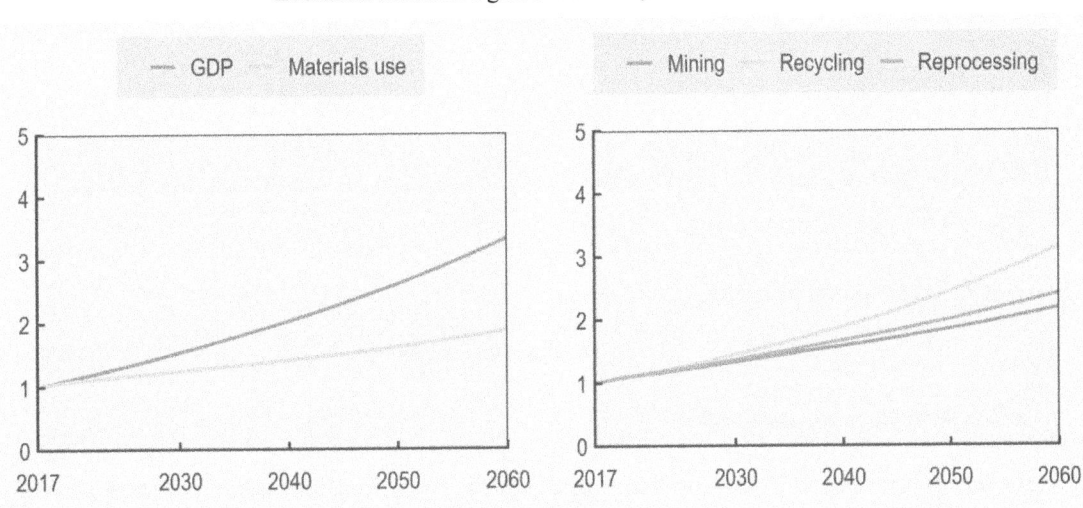

Note: The right-hand side panel shows the evolution of the sector output.
Source: OECD ENV-Linkages model.

StatLink https://doi.org/10.1787/888933884308

The increase in materials use applies to all material groups (biomass, fossil fuels, metals and non-metallic minerals) and all major regions in the world (OECD, BRIICS and Rest of the world) (Figure 1.4). Non-metallic minerals remain the largest materials group in the three regions. The rapid increases in the short run in the BRIICS support their strong short-term infrastructure growth (especially in China), and a more gradual ramp-up of infrastructure and minerals use in developing countries (the Rest of the world grouping).

Figure 1.4. Growth in materials use is projected for all regions

Materials use in Gt

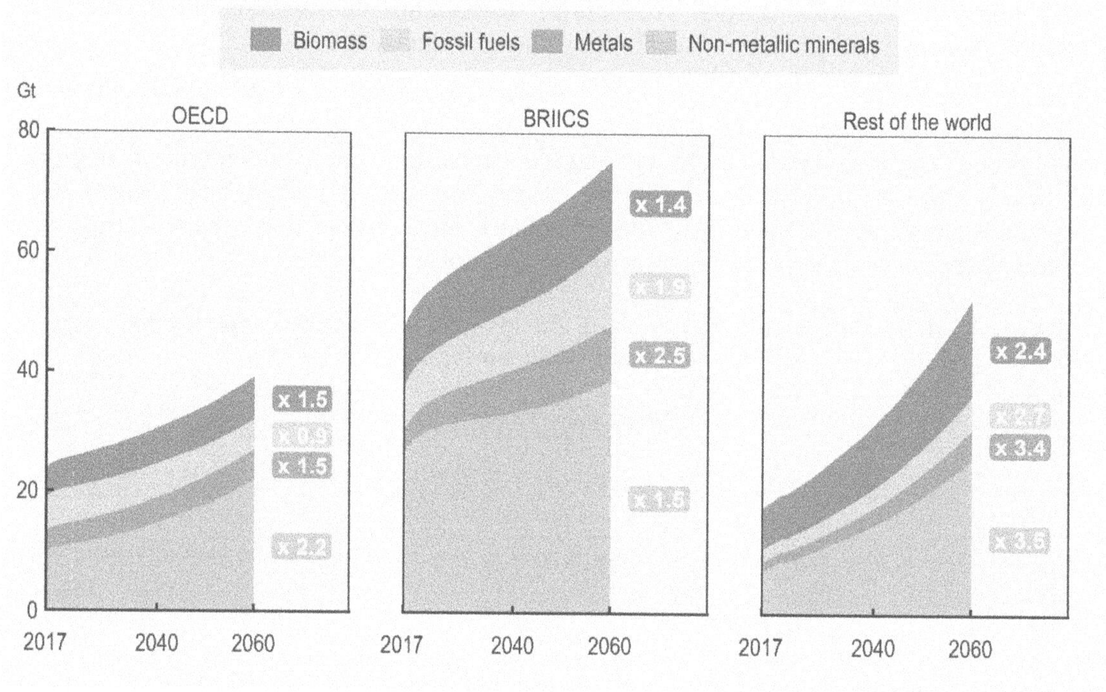

[Chart showing materials use (Biomass, Fossil fuels, Metals, Non-metallic minerals) in Gt for OECD, BRIICS, and Rest of the world regions from 2017 to 2060.

OECD: x1.5, x0.9, x1.5, x2.2
BRIICS: x1.4, x1.9, x2.5, x1.5
Rest of the world: x2.4, x2.7, x3.4, x3.6]

Source: OECD ENV-Linkages model.

StatLink https://doi.org/10.1787/888933884327

1.3. How is uncertainty accounted for?

There are many uncertainties surrounding the projections outlined above. These include assumptions about socioeconomic drivers and technology developments, as well as more systematic uncertainties, such as political stability or external shocks in the form of natural disasters and unforeseen climate feedbacks. Understanding the plausible range of future developments can provide policy makers with the information they need to implement policies that work regardless of the future economic context.

This report contributes to this by modelling the likely ranges for two key socioeconomic drivers: changes in population projections and income convergence. While other uncertainties – not least those directly affecting the competition between primary and secondary materials, and those affecting mining and recycling – could affect primary and secondary materials use projections more significantly, their quantification has been left for future research.

The range of materials use projections obtained for the varied assumptions about population growth and labour efficiency, which drives income convergence (see Section 3.4 in Chapter 3), are presented in Figure 1.5. With the links between growth, infrastructure and changes in the sectoral structure of the economy, the overall sectoral composition of the global economy also changes, causing differences between the different materials groups. Most prominently, biomass materials use is less sensitive to varying socioeconomic assumptions than the other material categories, as food is a basic commodity. Non-metallic minerals, and especially construction materials, are most closely linked to these socioeconomic assumptions.

Figure 1.5. Uncertainties on materials use is especially large for non-metallic minerals

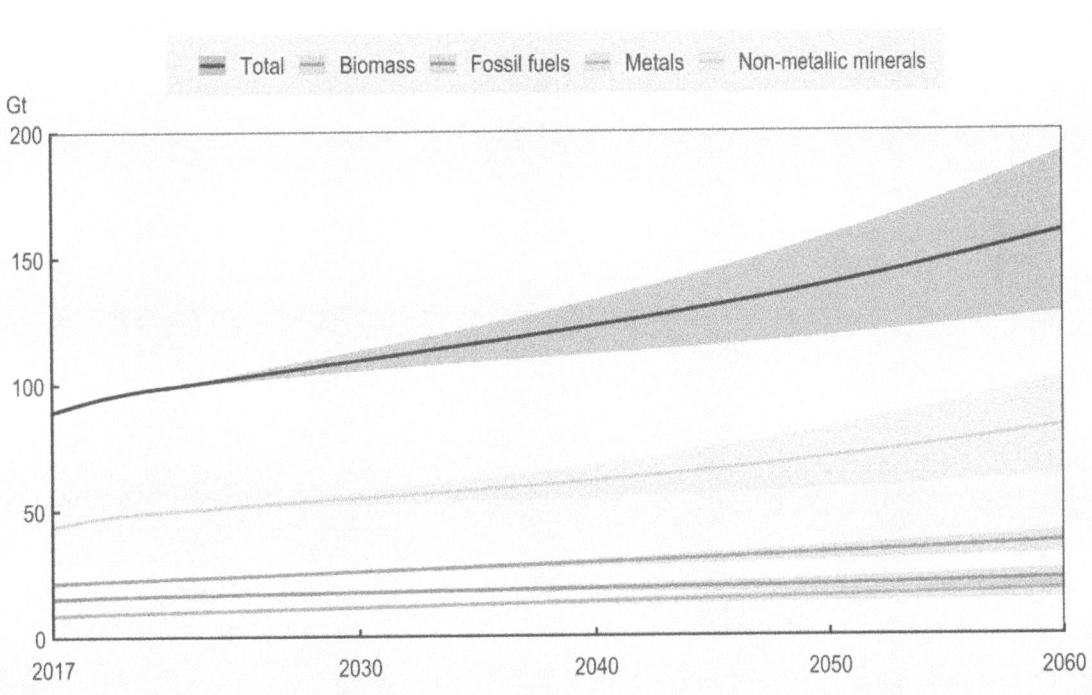

Source: OECD ENV-Linkages model.

StatLink https://doi.org/10.1787/888933884346

1.4. The projections imply large environmental and sustainable development challenges

The projected increase in materials use implies a significant increase in a wide range of environmental impacts, including acidification, climate change, eutrophication, land use, as well as water, human and terrestrial ecotoxicity.[3]

Most global environmental impacts are projected to at least double (Figure 1.6). Despite ongoing improvements in efficiency, and thus gradually declining environmental impacts per unit of production, declining ore grades and the increased scale of extraction and production of materials significantly worsen environmental impacts between now and 2060. Materials use also has significant consequences for climate change as most greenhouse gas emissions stem from materials-management sectors. The waste streams generated by current production and consumption patterns are also projected to increase. As a consequence, the future challenges linked to materials use affect waste management as well as the environment.

Figure 1.6. Global environmental impacts differ significantly across materials

Total environmental impacts in 2015 (lighter shaded area) and 2060 (full coloured area), index 1 for most polluting material in 2060

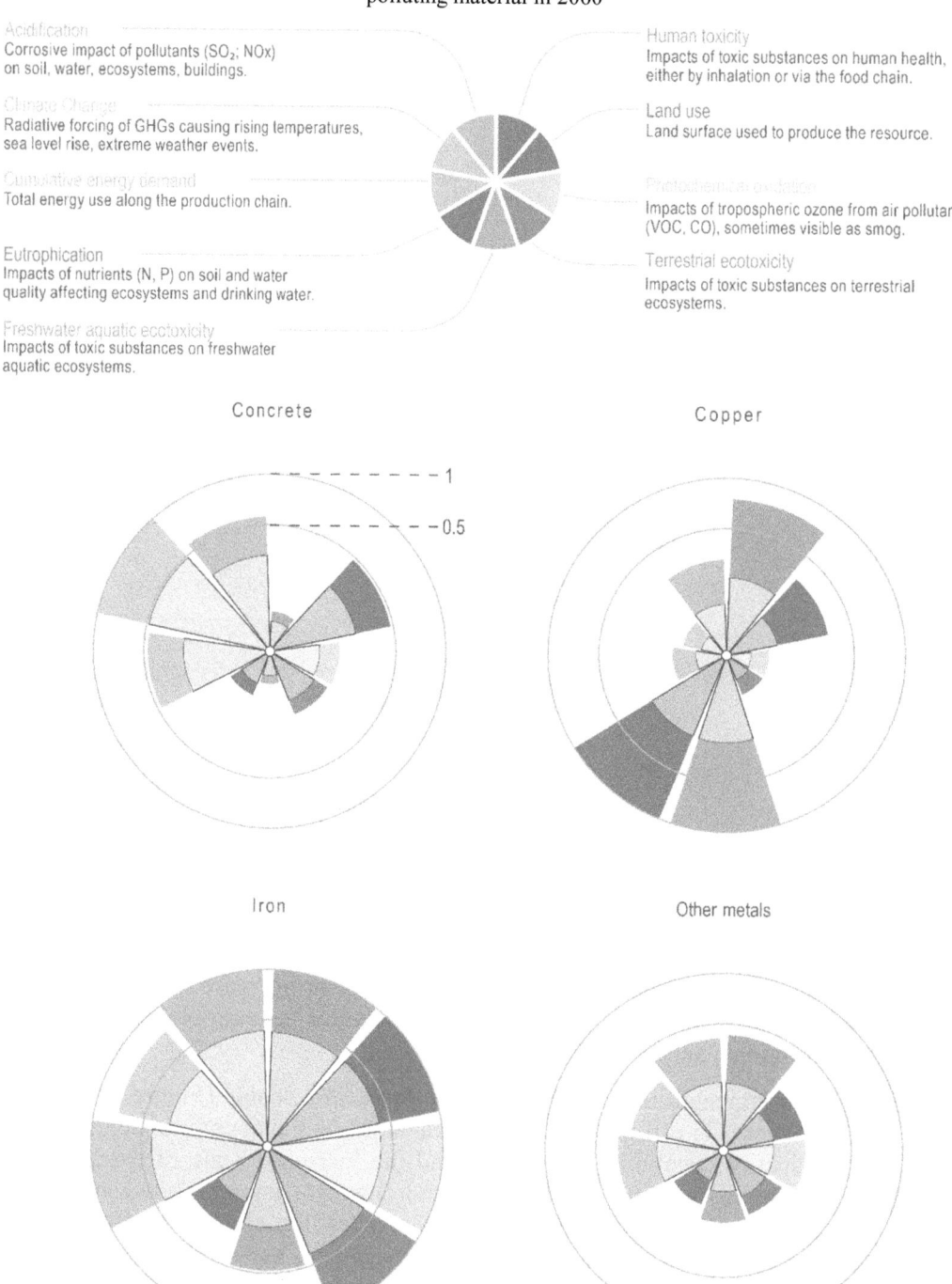

Note: Environmental impacts are presented for primary and secondary production combined. The lighter shading represents the value in 2015; the full coloured area reflects values in 2060. Impacts for "Other metals" reflect the combined impacts of aluminium, lead, manganese, nickel and zinc.
Source: OECD ENV-Linkages model.

StatLink https://doi.org/10.1787/888933884365

A large share of greenhouse gas (GHG) emissions is directly or indirectly linked to materials management. These come from the combustion of fossil fuels for energy, from agriculture, from manufacturing and from construction. The increased extraction and use of materials contributes to a global increase in GHG emissions, even if their contribution to overall emissions is projected to decrease relative to emissions not related to material management. Total emissions are projected to reach 75 Gt CO2-eq. by 2060 of which materials management would constitute approximately 50 Gt CO2-eq.

The ambitions of the Paris Agreement, including the Nationally Determined Contributions (NDC) and the "well below two degrees" objective, would not be met in the central baseline scenario. Additional policy efforts are required to meet these goals, such as including policies aimed at reducing emissions of GHGs in a comprehensive resource management policy package.

1.4.1. The links between economic activity, materials use and the Sustainable Development Goals are complex

Materials use, their economic drivers and their environmental consequences are central components of several Sustainable Development Goals (SDGs). The topic of materials use is represented most prominently in SDG 12, which aims to "ensure sustainable consumption and production patterns", explicitly targeting sustainable management and efficient use of natural resources (SDG 12.2).

The increase in materials use and the associated increase in environmental impacts projected in this report indicate some of the challenges and trade-offs to reaching the SDGs. While extraction and processing of primary materials leads to GDP growth and creates jobs (and thus contributes to SDG 8), these may not be sustainable, given their impact on the environment. Absolute decoupling of materials use and environmental degradation from GDP growth is desirable and targeted in SDG 8.4. Whilst this report shows that relative decoupling has occurred and is projected to continue to occur in the near future, overall materials use and the related environmental impacts are still projected to increase, putting SDGs 8 and 12 and other goals at risk.

The baseline projections show that renewable energy is projected to increase, supporting SDG 7, which encourages universal access to affordable, reliable, sustainable and modern energy. However, the expansion of renewable energy systems will also have implications for resource use; metals demand for wind- and solar power and battery systems is projected to increase. Furthermore, investment, infrastructure and construction relate to SDG 9 ("Build resilient infrastructure, promote inclusive and sustainable industrialisation and foster innovation"). Given the strong links between construction materials use and economic growth, which have been discussed as part of this report, the policy challenges in this domain are significant as it will be difficult to strike a balance between sustaining economic growth and limiting the environmental impacts caused by the use of non-metallic minerals in construction.

The significant toxic impacts on the environment of metals extraction and processing, as highlighted in Section 8.3, affects the sustainable use of terrestrial ecosystems (SDG 15), but through spillover effects also compromises SDG 14 on sustainable use of oceans and marine ecosystems.

Insights from this report and policies to stimulate resource efficiency and the transition to a circular economy, can also contribute more indirectly to achieving other Sustainable Development Goals. These include, but are not limited to, the links between biomass resources and ensuring sustainable food production systems (SDG 2.4), and the links between pollution caused by materials use (not least fossil fuels) and the objective to reduce health impacts from hazardous chemicals and air, water and soil pollution (SDG 3.9).

1.5. Improving materials efficiency will require better policies

Policy priorities should be determined considering the links between the use of a specific material and its economic drivers, as well as its impacts on the environment, and the criticality of its supply. The opportunities for substituting secondary for primary materials are also important in determining policy responses.

Macroeconomic indicators of materials productivity cloud the picture and obscure insights into what drives materials use. Countries at different levels of development use different material resources and have different opportunities to decouple materials use from economic growth. A granular approach is needed to understand which policy interventions may improve resource efficiency at the sectoral level, and how major environmental consequences can be avoided. An effective resource policy thus hinges on a detailed understanding of the economic drivers of materials use, and the environmental consequences.

Thus, non-metallic minerals (and steel), which are mainly used for construction, are closely linked to economic development and should be tailored to decoupling minerals use from infrastructure development, e.g. by stimulating the recycling of sand, gravel and concrete.

Policies tailored for metals will have to take into account the relatively high environmental impacts per kg of metals extraction and use; given the high projected growth rate of primary metals use and the stagnation of the share of secondary metals this issue is of rising importance over time.

Fossil fuel policies directly link in with reforming energy subsidies and climate change mitigation. Finally, policies for managing biomass resources are essential for achieving food security, ending hunger and avoiding the major bottlenecks in the land-water-energy nexus.

The objectives of resource efficiency and circular economy policies are varied and include increasing recycling, boosting economic growth, boosting employment and avoiding environmental impacts. This multitude of policy objectives requires a carefully balanced policy mix.[4] The OECD Policy Guidance on Resource Efficiency (OECD, 2016[3]) provides some generic recommendations (see Box 1.1).

To move towards more detailed and operational policy insights, quantification of the main linkages between economic activity, materials use and environmental pressures is needed. The complexities in the economic system are huge: sectoral economic activities are all connected, and value chains are increasingly global. Furthermore, there are complex links between the economic system and the environmental system. Together with the inherent uncertainty in future developments, these complexities imply the need for a systems approach. The global perspective and economy-wide assessment provided in this report provides a suitable reference point for identifying the scale of the problems associated with the increase in materials use under current policies and formulating the priorities for the policy response. A numerical assessment of the policies needed to transition to a more resource-efficient and circular economy can shed light on which policies may be most effective in reaching the various policy objectives.

Box 1.1. OECD Policy Guidance on Resource Efficiency

Resource efficiency policies can help to counteract current trends of continued material resource consumption and generate significant positive impacts for the economy and the environment.

Yet, to realise these benefits, these policies need to be further developed and mainstreamed. When designing policy, governments should focus on the following:

- Applying mixes of policy instruments that ensure a coherent set of incentives for resource efficiency along the product value chain.
- Implementing policies that promote resource efficiency across the lifecycle of products.
- Treating resource efficiency as an economic policy challenge and integrating it into cross-cutting and sectoral policies.
- Strengthening policy development and evaluation through better data and analysis.

As the globalisation of our economies continues and value chains stretch across multiple jurisdictions, there is also an increasing need for co-ordinated approaches at the international level. Governments should strengthen international co-operation, with particular focus on the following issues:

- Supporting businesses in their supply chain management efforts. As it is difficult for national governments to influence the way supply chains are managed due to their limited jurisdictional reach, this can be done more effectively at the international level.
- Alleviating barriers to trade and investment in environmental goods and services to ensure the diffusion of the best available environmental technologies.
- Mainstreaming resource efficiency into official development assistance more systematically.
- Harmonising environmental labels, information schemes and mutual recognition, reducing their duplication and associated costs across international markets.
- Improving resource efficiency data and indicators of resource efficiency challenges and policies, and making economic analysis of these more robust.

Source: OECD (2016), Policy Guidance on Resource Efficiency.

1.6. How is the report structured?

This first chapter has summarised the main findings of the modelling analysis. The remaining chapters are structured so as to facilitate the understanding of how the different economic drivers lead to changes in materials use, including primary and secondary materials, and to the environmental consequences linked with materials use (Figure 1.7).

Chapter 2 provides the context, purpose and methodology of the modelling analysis. Chapter 3 presents the economic baseline scenario, with a focus on the main economic drivers of materials use: socioeconomic trends, changes in demand patterns and in production processes. Focusing on the economic drivers presented in Chapter 3, Chapter 4 explains how these affect materials use. Chapter 5 then outlines the projections of primary materials use that emerge while Chapter 6 describes the projections for recycling and secondary materials use. Chapter 7 presents case studies that dive deeper into specific issues: projected copper stocks and flows, projected iron and steel stocks and flows, and projections for the criticality of materials. Chapter 8 discusses the environmental consequences of materials use, including a life cycle analysis of the environmental impacts of selected materials and an assessment of the links with climate change.

Figure 1.7. Roadmap of the report

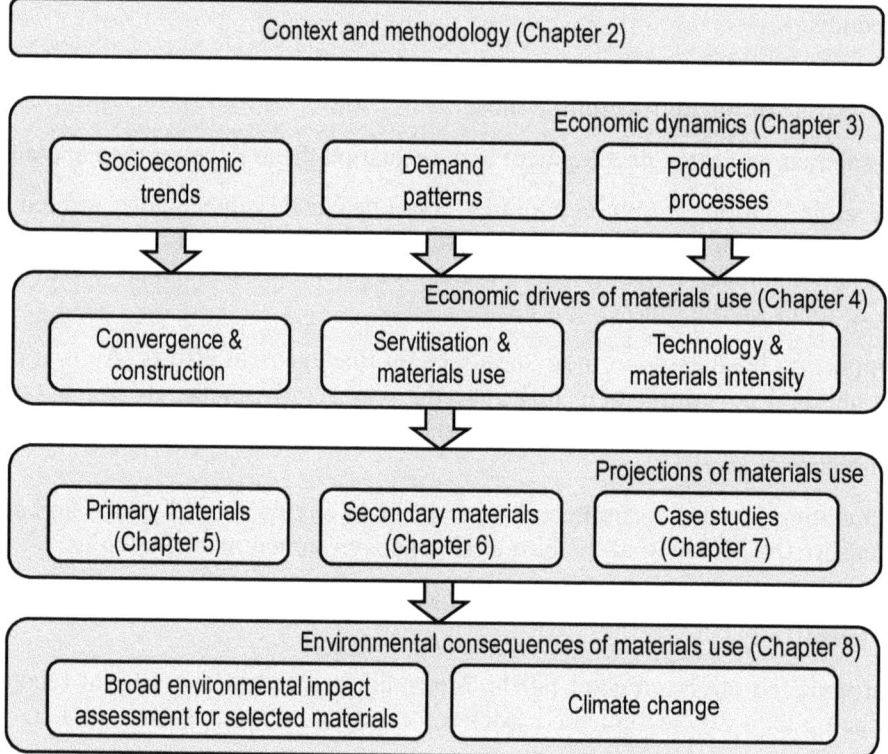

Notes

[1] In reality, individual economies do not experience the smooth pattern that the modelling framework portrays. External shocks, including natural disasters, internal and external political conflicts, are likely to disturb growth in specific countries in the short and medium run, with potential long-term consequences. These are projections, not predictions.

² Motor vehicles and electronics have low total materials intensities, but are relatively large users of metals, and so drive the fast increase in metals use.

³ The plausible range on the evolution of the studied environmental impacts is wider than that of global materials use, as these impacts are driven by specific dynamics for each material, region and sector. Furthermore, the differences between sectors and regions are large.

⁴ Existing modelling assessments of a resource efficient or circular economy transition have focused on a relatively small subset of enabling policies. This is largely due to the difficulty of implementing some types of "soft policies" – those that have an ambiguous effect on the economy – in a macroeconomic modelling framework (McCarthy, Dellink and Bibas, 2018[12]). The macroeconomic consequences of a broad policy mix to boost resource efficiency and transition to a circular economy remain therefore understudied.

References

Guillemette, Y. and D. Turner (2018), "The Long View: Scenarios for the World Economy to 2060", *OECD Economic Policy Papers*, No. 22, OECD Publishing, Paris, http://dx.doi.org/10.1787/b4f4e03e-en. [2]

McCarthy, A., R. Dellink and R. Bibas (2018), *The Macroeconomics of the Circular Economy Transition: A Critical Review of Modelling Approaches*, OECD Environment Working Papers, http://dx.doi.org/10.1787/af983f9a-en (accessed on 15 November 2017). [4]

OECD (2016), *Policy Guidance on Resource Efficiency*, OECD Publishing, Paris, http://dx.doi.org/10.1787/9789264257344-en. [3]

UN (2017), "World Population Prospects: key findings and advance tables", https://esa.un.org/unpd/wpp/publications/Files/WPP2017_KeyFindings.pdf (accessed on 18 May 2018). [1]

Part I.

The economic drivers of materials use

Chapter 2.

Modelling future materials use and its economic drivers

This chapter presents the rationale and policy background for this material resources outlook, including the methodology behind the projections of materials use to 2060. Specifically, it outlines the main economic model and explains how materials have been linked to the economic flows in the model. As the projections presented in this report reflect a baseline scenario, this chapter also explains the nature of such scenarios and the policy assumptions that are included in this particular scenario. The chapter concludes with a roadmap of the report to guide the reader through the various chapters and steps of the analysis.

2.1. Introduction

The economic growth of the last decades has been underpinned by the unprecedented use of natural resources. From 1980 to 2010, the use of global primary materials – comprising biomass, fossil fuels, metallic ores and non-metallic minerals – rose at about twice the rate of population growth, largely following the growth in Gross Domestic Product (GDP) (OECD, 2015[1]; Krausmann et al., 2009[2]).[1] The rate of growth was, not surprisingly, fastest in rapidly developing economies, while mature economies managed to partially decouple resource use from GDP growth (OECD, 2016[3]). But these trends appear insufficient to counteract the rising demand for materials-intensive commodities by a world population projected to reach more than 10 billion people by 2060 (UN, 2017[4]), all striving for higher living standards and increasing pressure on the environment (OECD, 2012[5]).

Materials extraction and use have important environmental implications. These include the pollution of local water sources, ecosystem disruption from mining facilities, the emissions of gases that contribute to climate change and air pollution, and environmental degradation from waste landfilling and incineration. This environmental damage in turn may have economic consequences, given that they affect human health, ecosystem services, labour productivity, capital and crop yields, for example.

As materials are essential for production processes, their availability is also fundamental for sustained economic growth. Previous OECD work (Coulomb et al., 2015[6]) shows that while absolute scarcity is not a major problem for most materials, risks to the security of supply of economically important materials – i.e. the criticality of materials use – can damage economic prospects. Absolute scarcity is thus not the most appropriate lens through which to study future materials use; focusing on future materials demand is more relevant.

Shifting away from unsustainable natural resource use would not only reduce environmental damages and supply risks, it could also create job opportunities, for example in collecting recyclables and preparing and processing secondary materials. While a large part of the changes in labour markets concerns a reallocation of employment between sectors, overall job gains are more likely to be achieved if the consequences of policies to achieve a more efficient use of resources are well managed to avoid skills mismatches and other adjustment problems.

These issues have led to increasing policy interest in promoting a transition to a more resource efficient and circular economy. Circular economy roadmaps now exist in the People's Republic of China (hereafter China, implemented in 2013), the European Union (2015), Finland, France, the Netherlands, and Scotland (2016), as well as Slovenia and Portugal (2017). Similar, but alternatively named, frameworks targeted at resource efficiency also exist in Japan (Fundamental Law for Establishing a Sound Material-Cycle) and the United States (The Sustainable Materials Management Program Strategic Plan). These policy frameworks tend to comprise one or more targets[2] along with a set of policies designed to achieve them (Box 2.2).

Effective policy action needs to be underpinned by a clear understanding of how economic trends affect materials use and the environment. This report aims to contribute to this by presenting an outlook of global materials use to 2060, examining in particular how socioeconomic trends drive changes in the use of different materials. The report focuses on three trends that drive materials use: (i) socioeconomic trends, which include population and income per capita growth in all countries (ii) changes in demand patterns, and (iii) changes in production processes (technical change).

The projections of materials use presented in this report are based on a dynamic general equilibrium modelling framework, where physical material flows are linked to specific economic activities. This modelling framework provides a systems perspective on the evolution of economic activity and material flows in different sectors and regions, with a focus on their interlinkages.

> **Box 2.1. Key definitions in the report**
>
> *Materials* in this report refer to the physical resources used in production and consumption. They include metals, non-metallic minerals, fossil fuels and biotic materials. An important material resource excluded in the analysis is water; as the water cycle involves more physical and climate processes, which are not modelled here. Land and air are also excluded.
>
> *Primary materials*, sometimes labelled as raw or virgin materials, refers to materials sourced from mining and extraction activities in their raw form, such as mineral ores. These materials are entering into the economic system for the first time.
>
> *Secondary materials* refers to materials that have already been used previously. A typical source of materials for secondary use is recycling.
>
> *Materials extraction* refers to the mass (physical weight) of primary materials extracted from the natural environment for use in the economy.
>
> *Recycling* refers to the process of converting waste into a supply of secondary materials that can then be reprocessed and reused. The processing of (recycled) secondary materials is labelled as *reprocessing*. For instance, the recycling of old cars produces scrap, which can be reprocessed into readily usable steel.
>
> *Materials use* is measured as the use in the economic system of material resources. There are two main ways to calculate regional materials use.
>
> 1. **Domestic material consumption** (DMC) measures the weight of the materials that are used in the domestic economic system (i.e. the direct apparent consumption of materials, excluding indirect flows). In economy–wide material flow accounting, DMC equals domestic extraction plus imports minus exports.
>
> 2. **Raw material consumption** (RMC) measures the weight of materials associated with domestic consumption, including materials embedded in internationally traded products (indirect use). RMC equals DMC plus indirect flows associated with imports minus indirect flows associated with exports. The defining difference between DMC and RMC is where in the economic process the physical materials are allocated. At the global economy-wide level, these indicators are equal.
>
> This report uses the DMC approach and applies it at the regional, sectoral level, i.e. it attributes materials use to the most directly related production activity.
>
> *Materials use* thus refers – in this report – to the sectoral flows of materials attributed to the first-use demand categories, e.g. the materials associated with the input of mining

> products in metal processing, not supply (extraction) of the material. The materials that are used can be domestically extracted or imported; but materials use does not account for indirectly associated flows.
>
> ***Materials intensity*** can be calculated as the ratio between the amount of materials used (typically in terms of weight) and the value of the related economic output.
>
> ***Resource productivity*** refers to "the effectiveness with which an economy or a production process is using natural resources" (OECD, 2015[11]); for material resources it is generally calculated as the inverse of materials intensity, i.e. expressed as dollars of output (or GDP) per tonne of material.
>
> *Source*: Based on OECD (2008[7]).

2.1.1. What is the value added of this report?

This report is the latest in an emerging body of modelling work which takes a demand approach to assess the economic consequences of improved resource efficiency and a circular economy transition. So far, much of this literature has focussed on individual countries and a small subset of policy instruments (Winning et al., 2017[8]; Cambridge Econometrics, 2014[9]). However, the POLFREE project took a global perspective, and examined a more comprehensive policy mix (Hu, Moghayer and Reynès, 2015[10]; Distelkamp and Meyer, 2016[11]). The United Nations Environment Programme's report Resource Efficiency: Potential and Economic Implications, produced by the International Resource Panel (IRP), is the most extensive assessment of future materials use to date (UNEP, 2017[12]). The report highlights how population and per-capita economic growth are expected to lead to a significant increase in materials use to 2050. As a result, the report projects that global materials use may increase from 84 Gigatonnes (Gt) in 2015 to 184 Gt in 2050. That report also highlights the potential to reduce materials use significantly through policies aimed at resource efficiency and climate change mitigation.

There are three major issues with the existing studies of future materials use. First, a recent assessment by the OECD showed that the results of many existing studies are strongly driven by assumptions about reduced materials use through technological change and changes in consumption trends, with only limited consideration of the costs involved in these evolutions (McCarthy, Dellink and Bibas, 2018[13]). Furthermore, this literature lacks projections on the evolution of secondary materials and the degree to which they can substitute primary materials, which is crucial for understanding the materials intensity and the degree of circularity of the economy. Finally, future scenarios are often presented without explicitly addressing uncertainties. This can give a misleading impression that future materials use can be accurately predicted when in fact future projections are inherently uncertain.

This report attempts to fill these gaps, at least partially. It takes a demand-based approach, focusing on the evolution of the economic drivers of materials use. Materials supply is limited by modelling increasing costs for expanding the supply of resources. Further, the modelling framework is used to provide scenario projections of how socioeconomic developments and technological change interact in driving both primary and secondary materials use. The modelling framework is also used to highlight the various sources of uncertainty and to model alterative scenarios.

This report focuses on the evolution of the use of primary and secondary materials. To do so, primary material extraction and secondary material recycling are modelled (see

Box 2.1 for definitions). Reuse and remanufacturing are not explicitly modelled, but are however included in the long-term trends in the evolution of consumption preferences.

The projections in this report are meant to shed light on the key mechanisms that link economic activity, materials use and environmental impacts. They present long-term trends, and cannot account for unforeseen future changes in climate, political systems, technology. They are thus not a predication of what will happen.

By presenting a detailed analysis of future materials use, their economic drivers and environmental consequences, this report can help policy makers understand the scale of the challenge in transitioning to sustainable resource use and the need for policy action. The baseline scenarios presented in this report also provide the backbone for assessing the economic consequences of policies aimed at improving resource efficiency and promoting the transition to a circular economy (Box 2.2).

Box 2.2. Policies for resource efficiency and a transition to a circular economy

The policies to promote a transition to a more resource efficient and circular economy are very diverse. The policies that are most frequently referred to include green public procurement, support for research and development (R&D), extended producer responsibility (EPR) schemes, product design standards, recycling rate standards, waste disposal taxes, and consumer education campaigns. Some of these policies have existed for many years, but are being discussed with a view to strengthen them. In other cases, there is an intention to introduce policies in the near future.

While several governments have set up a policy framework to achieve a transition to a circular economy, circularity is presented as a means to achieve government objectives rather than a goal in itself. Similarly, resource efficiency policies do not aim at reducing the use of resources per se, which is generally a source of economic growth, but instead they address the market failures that are related to the use of resources. For example, a tax on fossil fuel use directly applies to the material, but only indirectly to the relevant market failure (i.e. the exclusion of costs related to climate change and air pollution caused by fuel emissions).

A previous OECD report, *The Macroeconomics of the Circular Economy Transition*, described three different layers of circularity, with increasingly broad coverage: (i) closing resource loops; (ii) slowing resource loops; and (iii) narrowing resource loops (McCarthy, Dellink and Bibas, 2018[13]). All these explicitly or implicitly aim at addressing the market failures associated with materials use, i.e. the failure to account for externalities in resource pricing, e.g. of the failure to include the social cost of carbon in extracting or processing materials; of the failure to address local environmental consequences associated with extraction; or the failure to include the environmental externalities associated with waste generation. Furthermore, there are economic inefficiencies associated with the inefficient use of scarce resources. The extent to which policy can close, slow or narrow resource loops differs by type of instrument.

2.2. How is future materials use modelled?

The economic and material projections presented in this report are based on a suite of global modelling tools, as described in Figure 2.1 (see Annex 2.A for further details). The models are used to project sectoral and regional economic activities and materials use over the medium and long term – up to 2060. The OECD ENV-Growth model, a global macroeconomic growth model, projects macroeconomic trends by country. The projections of ENV-Growth are then combined with assumptions on structural change in the second step to calibrate the ENV-Linkages model, a global dynamic computable general equilibrium (CGE) model (Chateau, Dellink and Lanzi, 2014[14]). ENV-Linkages details economic activities linked between sectors and across regions.[3] ENV-Linkages, which is the core tool of the analysis, has been enhanced to include materials use and recycling activities, so as to obtain scenarios not only of economic activities, but also of materials use in physical terms for both primary and secondary materials.

Figure 2.1. The scenario projections build on a complex modelling framework

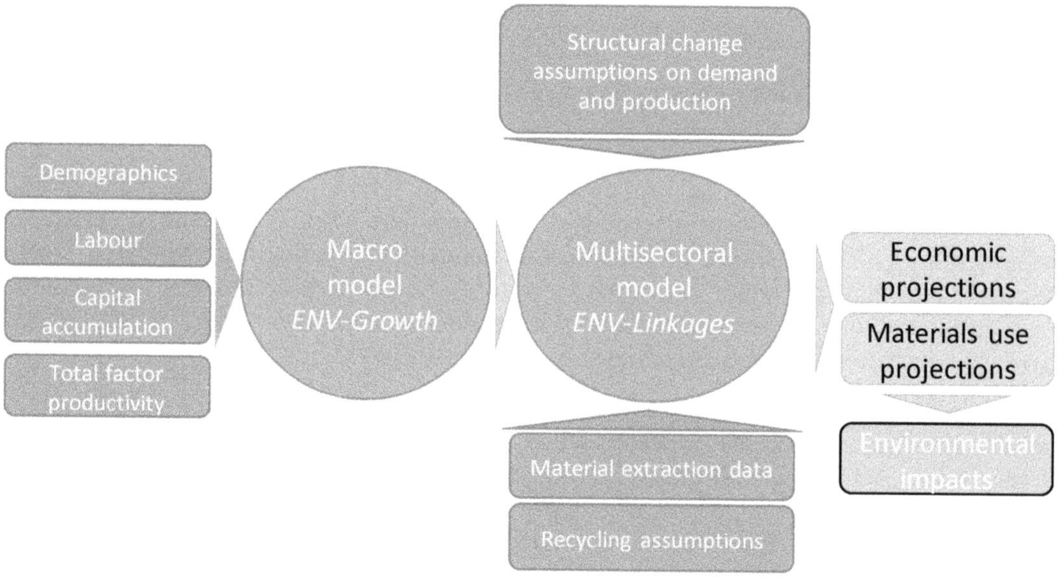

The first step consists of projecting GDP on the basis of macroeconomic trends for employment, Total Factor Productivity (TFP), capital accumulation, and current account balances for 230 individual states, using the ENV-Growth model. The ENV-Growth model is based on the basic idea of "conditional convergence" across countries, as explained in e.g. (Guillemette and Turner, 2018[15]). In this conditional convergence set up, countries' income levels per capita converge towards their country-specific long-term potential (their 'frontier'). As a result, lower income countries tend to grow faster than more mature economies, since a large part of the current income differences can be explained by the distance of the country to its domestic frontier.. ENV-Growth adds projections for developing countries to the country coverage of the Long Term Model of the OECD's Economics Department (Guillemette and Turner, 2018[15]), which is used to produce the official OECD country-specific long-term macroeconomic projections for OECD countries and non-OECD G20 countries. For these 46 countries, the macroeconomics trends of both models are identical, at least for the central baseline scenario.

The second step uses the ENV-Linkages model to enrich the macroeconomic projection with sectoral projections and interlinkages between sectors, final demand patterns and remuneration of primary factors labour, capital, land and natural resources. ENV-Linkages is a multi-sectoral, multi-regional model that links economic activities to energy and environmental issues. It is based on the Social Accounting Matrices (SAMs) and economic data contained in the Global Trade Analysis Project (GTAP) database.[4] The version used for the current analysis contains 45 economic commodities (Table 2.A.1 in Annex 2.A) and 25 regions (described in Table 2.1), including trade flows.[5]

Table 2.1. ENV-Linkages model regions

Macro regions		ENV-Linkages countries and regions	Most important comprising countries and territories
OECD	OECD America	Canada	Canada
		Chile	Chile
		Mexico	Mexico
		USA	United States of America
	OECD Europe	OECD EU 17	Austria, Belgium, Czech Republic, Denmark, Estonia, Finland, Greece, Hungary, Ireland, Luxembourg, Netherlands, Poland, Portugal, Slovak Republic, Slovenia, Spain, Sweden
		OECD EU 4	France, Germany, Italy, United Kingdom
		Other OECD Eurasia	Iceland, Israel[1], Norway, Switzerland, Turkey
	OECD Pacific	Australia and New-Zealand	Australia, New-Zealand
		Japan	Japan
		Korea	Korea
Non OECD	Other America	Brazil	Brazil
		Other Latin America	Other non-OECD Latin American and Caribbean countries
	Eurasia	Caspian region	Armenia, Azerbaijan, Georgia, Kazakhstan, Kyrgyzstan, Tajikistan, Turkmenistan, Uzbekistan
		Other EU	Bulgaria, Croatia, Cyprus[2], Latvia, Lithuania[3], Malta, Romania
		Other Europe	Albania, Andorra, Belarus, Bosnia and Herzegovina, Gibraltar, Former Yugoslav Rep. of Macedonia, Rep. of Moldova, Montenegro, San Marino, Serbia, Ukraine
		Russia	Russian Federation
	Middle East and Africa	Middle East	Bahrain, Iraq, Islamic Rep. of Iran, Kuwait, Lebanon, Oman, Qatar, Saudi Arabia, United Arab Emirates, Syrian Arab Rep., Yemen
		North Africa	Algeria, Egypt, Libya, Morocco, Tunisia, Western Sahara
		Other Africa	Sub-Saharan Africa excl. South Africa
		South Africa	South Africa
	Other Asia	China	People's Rep. of China, Hong-Kong (China)
		India	India
		Indonesia	Indonesia
		Other ASEAN	Brunei Darussalam, Cambodia, Lao People's Dem. Rep., Malaysia, Myanmar, Philippines, Singapore, Thailand, Timor-Leste, Viet Nam
		Other non-OECD Asia	Other non-OECD Asian and Pacific countries

Notes:
[1] The statistical data for Israel are supplied by and under the responsibility of the relevant Israeli authorities. The use of such data by the OECD is without prejudice to the status of the Golan Heights, East Jerusalem and Israeli settlements in the West Bank under the terms of international law.

[2] Note by Turkey: The information in this document with reference to "Cyprus" relates to the southern part of the Island. There is no single authority representing both Turkish and Greek Cypriot people on the Island. Turkey recognises the Turkish Republic of Northern Cyprus (TRNC). Until a lasting and equitable solution is found within the context of the United Nations, Turkey shall preserve its position concerning the "Cyprus issue".

Note by all the European Union Member States of the OECD and the European Union: The Republic of Cyprus is recognised by all members of the United Nations with the exception of Turkey. The information in this document relates to the area under the effective control of the Government of the Republic of Cyprus.

[3] Lithuania has become member of the OECD in July 2018. The regional aggregation of the model could not be revised to reflect this.

The model describes capital accumulation using capital vintages, in which technological advances trickle down slowly over time as capital stocks get upgraded and renewed. The structure of the model allows for a detailed assessment of the environmental consequences of economic growth and of specific changes at regional and sectoral levels. In particular, it links economic activity to materials use, and to indicators of environmental pressure, such as emissions of greenhouses gases (GHGs) and outdoor air pollutants.

For this report, the ENV-Linkages model has been enhanced to include materials use projections. Material flows are linked to the economic flows at the detailed sectoral level. The basic principle for linking is that each physical flow (materials use in tonnes) is attached to the corresponding economic flow (materials demand in dollars). A coefficient of physical use per dollar of demand is calculated and used to project materials use to 2060. This linking procedure is explained in detail in the next section.

2.2.1. How are material flows integrated into the modelling?

Including projections of material flows in an economic modelling framework requires substantial changes to the model, especially compared to other approaches that focus only on the physical material flows. More specifically, including materials use in ENV-Linkages required three main steps: (i) including the extraction of primary materials in the model, (ii) modelling recycling and secondary material processing, and (iii) modelling the substitution between primary and secondary materials, when possible. This section first explains the reasons for choosing an economic modelling framework, and then details the three steps used to include material flows in the model.

The chosen CGE modelling framework is well suited to representing the economic drivers of material flows. While other approaches to tracking materials such as Material Flow Analysis (MFA) quantify flows and stocks of materials in terms of physical flows, across their lifetime, they usually exclude the economic dimension. The CGE approach, on the other hand, tracks material flows in the economy at the detailed level of individual materials in specific sectors and by region, all embedded into a global systemic approach where economic activities are linked across sectors and regions.

Primary material extraction is linked to demand

Existing OECD data on materials does not have sufficient regional and sectoral coverage to allow a full global assessment of materials use (OECD, 2015[1]). The latest UNEP dataset on physical material flows (UNEP, 2017) is therefore used as the basis for the projection of primary material extraction.[6] The detailed regional and material coverage of this dataset, for all GTAP regions, as well as for 60 materials, allows the physical material flows to be linked to the economic flows in the ENV-Linkages model.

The technical difficulty in linking the available material flow data to the economic data, in order to be able to make projections of material extraction for the future, lies in the mismatch in terms of sectoral aggregation. Indeed, the economic data included in the underlying Social Accounting Matrices and economic databases, based on the GTAP database, comprise the extraction of various materials into an aggregated production sector (labelled mining).[7] The mining sector includes extraction of metallic and non-metallic mineral ores, but excludes fossil fuel extraction. This implies that the link between material flows (which are available at the level of individual materials) and economic flows can only be made in an aggregated manner, as insufficient information is available to identify the economic flows for each material separately.

When the economic extraction activity is modelled in sufficient detail in the underlying economic database (GTAP). A direct mapping of specific material flows to the associated extractive activity can be made. Material flows are then directly linked to the economic output of the corresponding extraction sector. For instance, the use of coal can be linked to coal mining, and sugar crops to sugar cane and beet production. As described in Table 2.2, biomass and fossil fuel resources can be modelled like this.

Table 2.2. Material flows linked to sectoral output

Category	Sector	Material
Fossil fuels	Coal extraction	Anthracite
		Other Bituminous Coal
		Peat
	Gas extraction	Natural gas
		Natural gas liquids
	Oil extraction	Crude oil
		Oil shale and tar sands
Biomass resources	Livestock (cows, other)	Grazed biomass
		Other crop residues (sugar and fodder beet leaves etc.)
		Straw
	Sugar cane, sugar beet	Sugar crops
	Forestry	Timber (Industrial round wood)
		Wood fuel and other extraction
	Fishing	All other aquatic animals
		Aquatic plants
		Wild fish catch
	Vegetables, fruit, nuts	Fruits
		Nuts
		Vegetables
	Oil seeds	Oil bearing crops
	Plant-based fibres	Fibres
	Wheat	Wheat
	Paddy rice	Rice
	Cereal grains n.e.c.	Cereals n.e.c.
	Crops n.e.c.	Other crops n.e.c.
		Pulses
		Roots and tubers
		Spice - beverage - pharmaceutical crops
		Tobacco

Source: Own assumptions.

When such as direct mapping of materials use to the output of an economic sector is not possible, the economic activity of the downstream processing sectors which demand the extracted commodities are used to project materials use. In this case, an additional step is needed to link material flows to the input of the extractive activity (e.g. mining) into the relevant processing sector, as described in Table 2.3. For each sector, the modelling framework describes the economic flows from one sector to another (in the input-output table). For example, this means that the physical use of iron ores is linked to the demand for mining products by the iron and steel sector.

Table 2.3. Material flows linked to inputs into processing sector

Category	Input from this extraction sector	Into this processing sector	Materials
Fossil fuel	Coal extraction	Coal power	Lignite (brown coal)
			Other Sub-Bituminous Coal
		Iron and steel	Coking Coal
Non-metallic minerals *	Non-metallic Minerals*	Construction	Gypsum
			Limestone
			Sand gravel and crushed rock for construction
			Structural clays
		Construction	Ornamental or building stone
		Chemicals, rubber, plastics	Chemical minerals n.e.c.
			Fertiliser minerals n.e.c.
			Salt
			Chalk
			Dolomite
		Non-metallic minerals	Industrial minerals n.e.c.
			Industrial sand and gravel
			Other non-metallic minerals n.e.c.
			Specialty clays
Metals	Mining (other than fossil fuels)	Iron and steel	Iron ores
		Non-ferrous metals	Bauxite and other aluminium ores
			Chromium ores
			Copper ores
			Gold ores
			Lead ores
			Manganese ores
			Nickel ores
			Other metal ores
			Platinum group metal ores
			Silver ores
			Tin ores
			Titanium ores
			Zinc ores
		Refining	Uranium ores

Note: * The non-metallic minerals sector is not an extraction sector, but the assumption is made here that construction materials that need to be processed (e.g., cement) follow the economic flow of the non-metallic minerals processing sector into construction rather than the mining sector into non-metallic minerals.
Source: Own assumptions.

The advantage of the demand-based approach is that the model is able to capture the effects of structural changes as well as modifications of consumption patterns, which drive changes in materials demand. For instance, the consumption of wood for construction, paper, furniture and other wood products drives the extraction activity for the forestry sector. An added advantage of this modelling framework is the inclusion of explicit trade flows between countries which allow material trade flows, as well as materials embedded in finished products, to be tracked. Tracking material trade flows sheds light on countries' *raw material consumption*, which helps diagnose their material footprint. For example, metallic objects which are consumed in Europe or the USA, but made in China with Australian metal ores can be properly accounted for. Furthermore, this helps understand the dynamics of final demand in order to design successful policies to reduce the environmental impacts of materials use along the chain of production, consumption and disposal. Projections of future raw material consumption levels in different countries are left for future studies.

How are recycling and secondary material processing included?

The general idea for modelling the recycling sector in ENV-Linkages is to formally introduce the recycling and secondary processing sectors into the national accounts. The former represents the activity which transforms waste and scrap into secondary raw materials (materials that can then be reprocessed into new materials). The latter takes the scrap from the recycling sector and reprocesses it into new materials.

The process involved is described in Figure 2.2. The first step is to separate the GTAP database's Other manufacturing sector into manufacturing and recycling. This is done by relying on the Exiobase database (Stadler et al., 2018[16]), which distinguishes recycling sectors. Thus, the sectors in Exiobase are used to split the GTAP sector and to create a recycling sector in ENV-Linkages.[8]

Figure 2.2. Splitting the recycling and processing sectors provides relevant detail

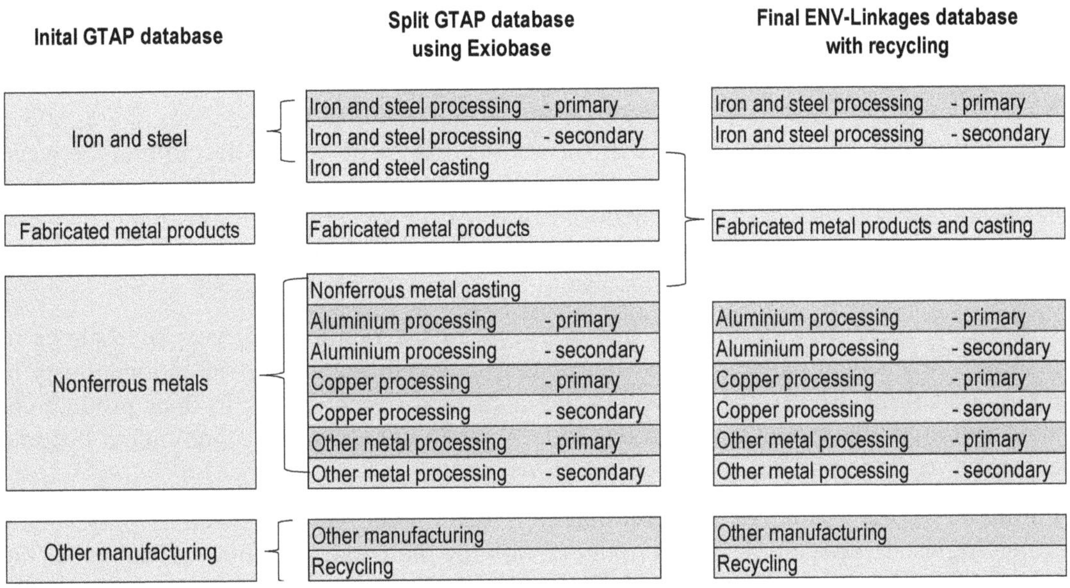

Inital GTAP database	Split GTAP database using Exiobase	Final ENV-Linkages database with recycling
Iron and steel	Iron and steel processing - primary Iron and steel processing - secondary Iron and steel casting	Iron and steel processing - primary Iron and steel processing - secondary
Fabricated metal products	Fabricated metal products	Fabricated metal products and casting
Nonferrous metals	Nonferrous metal casting Aluminium processing - primary Aluminium processing - secondary Copper processing - primary Copper processing - secondary Other metal processing - primary Other metal processing - secondary	Aluminium processing - primary Aluminium processing - secondary Copper processing - primary Copper processing - secondary Other metal processing - primary Other metal processing - secondary
Other manufacturing	Other manufacturing Recycling	Other manufacturing Recycling

The metal reprocessing sectors are disaggregated to distinguish between primary and secondary production for six metals and metal groups (Figure 2.2).[9] The iron and steel sector in GTAP is split into 'Iron and steel processing – primary', 'Iron and steel processing – secondary', as well as 'Iron and steel casting'. In a similar spirit, the non-ferrous metal sector in GTAP is split into 11 sub-sectors: primary and secondary processing for aluminium, copper, and other non-ferrous metals, as well as a common casting sector. While processing is of interest in this study, the casting sectors are not as relevant, so both iron and steel casting and non-ferrous metal casting are aggregated with the downstream use into a new fabricated metal products and casting sector.[10]

How is substitution between primary and secondary materials modelled?

Industries can produce commodities using primary or secondary materials as inputs, based on the level of quality desired for the application and the relative prices of the primary and secondary input materials. Primary materials tend to be of higher quality. As a consequence, primary and secondary materials are deemed highly substitutable but are not perfect substitutes. In the ENV-Linkages model, both primary and secondary processing sectors represent activities that compete to provide the same commodity: refined metal. Thus, while 12 processing sectors have been added, only 6 new commodities are produced. To reflect this in the model, primary and secondary processing are represented as competing technologies. Both technologies are characterized by capital, labour and energy costs, as well as material inputs. Furthermore, primary and secondary technologies produce a very similar good (the assumption is that primary and secondary materials are very good substitutes). The proportion of primary and secondary materials processing results from a price-based competition. . The elasticity of substitution between both processes is limited, as the outputs of primary and secondary materials processing are not always identical, e.g. in quality terms, and to account for other barriers – such as local preferences – that limit the switch in demand for the two processes. Mannaerts (2004[17]) suggests elasticities of 3 for steel, 2 for aluminium, as well as 2 for plastic and 4 for paper. The elasticities of substitution between primary and secondary minerals in the central baseline scenario are set to 2 for all materials.

2.2.2. The model accounts for decoupling of materials use from economic growth

The projections of materials use depend on the extent to which decoupling between economic growth and materials use will take place. Materials productivity increases when economic activity rises faster than materials use. At the sectoral level, this happens when more output is generated with the same amount of material inputs. Box 2.3 explains the different types of decoupling that can occur.

Historically, economic decoupling has been one of the dominant drivers of changes in materials intensity at the sectoral level. Economic decoupling is handled endogenously in the model: firms are assumed to weigh the costs of different inputs in their production, and minimise their costs of production. If material saving technologies become competitive, firms will decrease their materials inputs.

Furthermore, the timing of the decoupling is determined by the speed at which new technologies are adopted in the economy (following the creation of new capacities or the retrofit of existing ones). The ENV-Linkages model thus distinguishes between new and old production technologies. This "capital vintage" structure drives economic decoupling in the model as over time the old technologies become obsolete and newer and more efficient technologies are used.

Similarly, households make decisions on how to spend their income in order to maximise the utility from consumption based on the relative prices of consumption goods. As all economic sectors are linked in the economic model, a shift in the price of one commodity will induce an adjustment on all markets until relative prices, demand and supply are aligned again. Economic decoupling can also come from (total) factor productivity improvements. These improvements imply that the same volume of output can be produced while decreasing the volume of inputs, including material inputs.

Changes in the physical intensity of materials use at the sectoral level can in principle also play a role in projecting future materials trends. In this report, the physical materials intensity of the relevant economic activities – the output of the extractive sector when sufficient detail is available, and the input of the extractive sector in the processing sector when extraction is modelled in less detail – is kept constant. This is a common assumption in economic models that include resource use, such as Hu et al. (2015[10]), Meyer et al. (2015[18]), and UNEP (2017[12]). Thus, no physical decoupling is assumed in the baseline scenario and materials use in physical terms scales with the corresponding economic flow. This implies that all decoupling effects are captured through the economic framework.

Box 2.3. Understanding decoupling

The term "decoupling" is used to describe an improvement in resource efficiency, usually at the aggregate level of an economy. *Relative decoupling* refers to a situation in which the value of economic output and the amount of resource inputs are both growing, but the former is growing faster than the latter. This phenomenon was observed in the global economy during the last decade of the 20th century (OECD, 2016[3]). *Absolute decoupling* refers to a situation in which the value of economic output is growing while the amount of resource inputs used is shrinking. There is little evidence for absolute decoupling occurring in any country once the materials embodied in intermediate imports are taken into account (e.g. (Wiedmann et al., 2015[19]; OECD, 2016[3])). The last 15 years do not show a relative decoupling at the global level, since global material extraction grew faster than global GDP (Giljum, 2014[20]; Schandl, 2017[21]).

A different typology of decoupling is relevant at the sectoral level. The degree of *economic decoupling* reflects the level of reliance of key manufacturing sectors (e.g. construction) on intermediate inputs from the extractive sectors (i.e. mining). For example, the change in the amount of mining input in the iron and steel sector compared to other production inputs. As economies evolve, they may be able to rely less on mining inputs and more on other inputs, such as services. Furthermore, as factor productivity increases, the output of these manufacturing sectors may grow without increasing the use of (material) inputs. Across sectors, economic decoupling also results from a change in the structure of the economy towards less materials-intensive sectors, e.g. a shift away from industry towards services.

In addition to economic decoupling, there can be changes in the material content of mining products feeding into the manufacturing sectors. For example, manufacturing could switch to higher quality, more expensive but lighter metal sheets in production. This *physical decoupling* reflects changes in the ratio of the amount of materials (in weight terms) per unit of value of the associated economic flow (in dollar terms). Changes in the material content of inputs can also imply physical recoupling: one example would be the increase in volume of iron ore used by the iron and steel sector to compensate for declining ore grades.

2.2.3. Material reserves, stocks and supply constraints are not easily integrated

By focusing on the use of materials, there is a risk that the projections imply a demand for materials that cannot be met by the existing supply. Ideally, an economic analysis of material flows over time would track the evolution of the stocks of materials. Two types of stocks need to be differentiated:

1. Below ground stocks (i.e. material reserves which can be extracted): these are the source of primary materials. The availability and accessibility of reserves restricts mining activities and thus the supply of primary materials.

2. Above ground stocks of materials (i.e., materials embedded in products currently in use): these form the future supply of secondary materials. The supply of secondary materials depends on the availability of recyclable materials in the waste stream. The size of the waste stream in turn depends on the historical material stocks of materials embedded in single use products (e.g., food packaging), semi-durable goods (e.g., cars or phones), or even durable goods (e.g., buildings and infrastructure). Waste recyclability depends on a number of factors such as material concentration, contamination, or design.

The lengths of the life cycle of above-ground stocks vary widely for different materials embedded in different products. Materials used in short-lived consumer products have a much shorter life than materials embedded in houses and roads. In developed economies where a large share of investments is in replacement of existing products and infrastructure, one may expect stable stocks sooner or later. Emerging and developing economies are projected to continue to build up their stocks for the next decades.

Unfortunately, material stock accounting data are not readily available. The representation of (semi-)durable goods in large-scale economic models is even more problematic. Given the complexity of the task and lack of data, the analysis does not track. The demand for materials projected in this study therefore cannot be directly compared with the availability of supply as stocks fluctuate over time. Furthermore, the analysis of stagnation of stocks (so-called saturation effects) can only be done at the level of flows, not for per-capita stocks; Section 3.4 discusses these potential saturation effects in more detail. Case studies in Chapter 7 further explore the role of stock dynamics in the evolution of primary and secondary metal use for copper and iron and steel.

Alternatively, one could add a fixed supply constraint on either primary or secondary materials. But this is not in line with basic economic non-renewable resource theory, which states that the availability of scarce resources is not constant, but is influenced by technological developments, exploration of new reserves, etc. As an example, Figure 2.3 highlights how estimations of reserve life for selected metals – i.e. the amount of years that demand can be satisfied by currently known reserves –have been revised over time. As a result, current estimates are in many cases as large as those made in the 1940s and 1950s (Panels A and B). Figure 2.3 also shows that that while material prices fluctuate over time, long-term prices do not increase dramatically, indicating that scarcity does not seem to be a strong concern currently (Panel C). As a result, restrictions imposed by limited reserves and stocks are mimicked in this report by putting constraints on the supply of natural resources and allowing them to evolve over time.

Despite the lack of stock accounting, a dynamic analysis of material flows over time can still provide interesting insights into the evolution of materials use. For instance, the changing geographical patterns of economic growth imply that peak periods may appear in different regions in different decades, triggering peak periods of demand for materials.

Figure 2.3. Historic estimates of key indicators indicate no absolute supply scarcity for selected metals

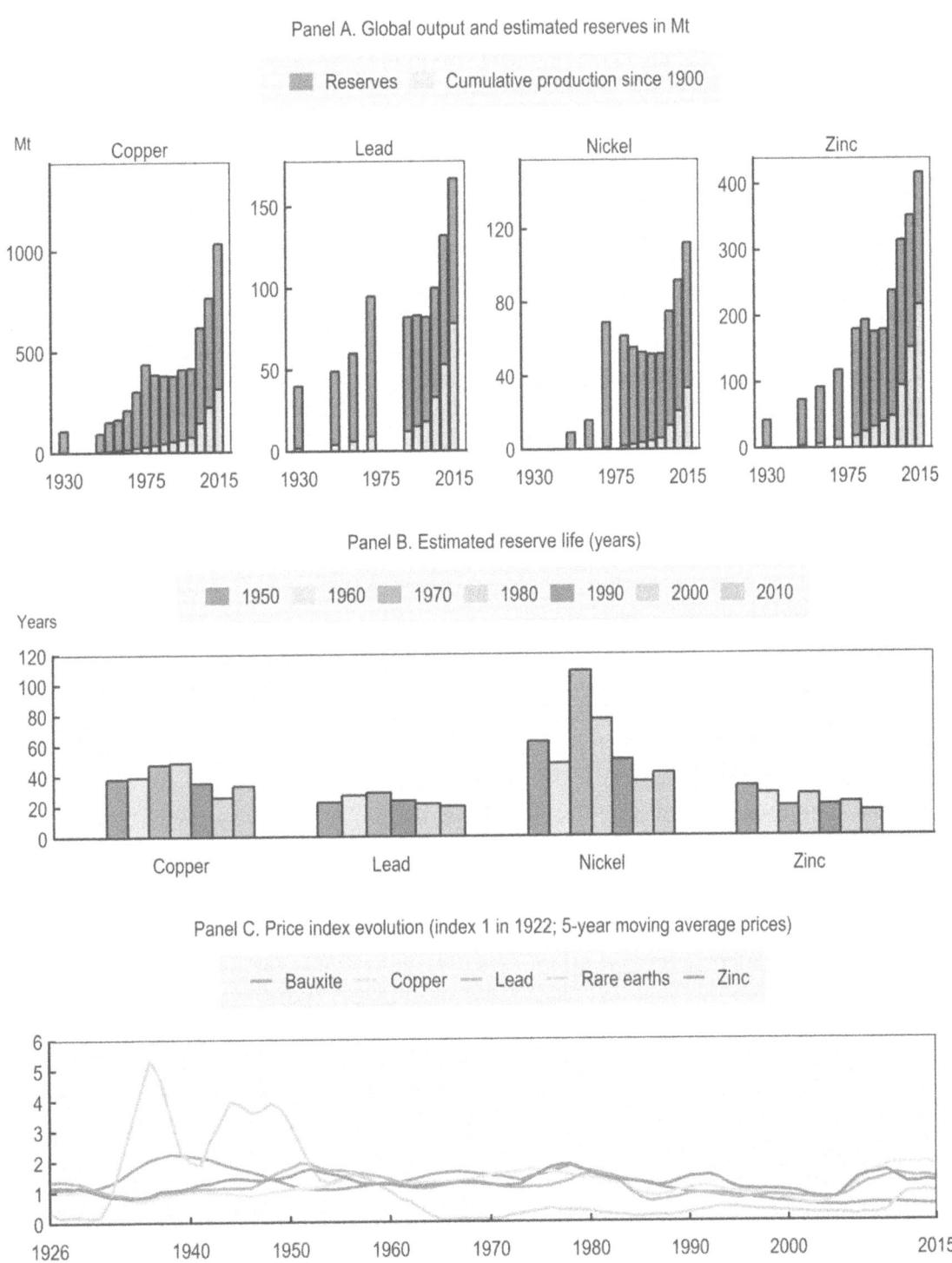

Source: Panel A: Crowson (2011[22]); Panel B: USGS (2017[23]), reproduced from Humphreys (2011[24]); Panel C: Annual Average U.S. Producer Prices, converted using Consumer Price Index conversion factor, with 1998 as the base year (Kelly, 2014[25]).

2.3. What is the baseline scenario approach?

This report uses a baseline scenario approach to explore future materials use as linked to economic activity. It is a scenario approach as it describes an internally consistent set of trend developments of all economic and environmental variables of the model. By excluding the analysis of the consequences of changes in relevant policies, the scenario describes trends for current policies, i.e. a "business-as-usual" baseline projection.

A baseline scenario projection describes one possible future development – it is not a prediction of future developments. Rather, a baseline scenario projection is calibrated so as to reproduce past and expected trends for several key economic and environmental variables, including demographic trends, urbanisation and globalisation trends.

Reproducing past trends does not mean that the central baseline scenario is a linear extrapolation into the future of recent trends in economic activity. Rather, the trends in the underlying drivers of economic growth are modelled, and careful assumptions are made about their plausible evolution in the coming decades. This evolution is not linear, as the trends in the drivers are not linear, and as the economic systems responds to changes in the drivers in a non-linear way.[11] For example, the effects of current policies tend to fade over time. Furthermore, preferences are assumed to gradually converge across countries, as are productivity levels – but with respect to country-specific constraints and technology frontiers. The general equilibrium model brings all these non-linear trends together into an internally consistent set of developments of all model variables. At the global level, this is a closed system: global exports equal global imports, global savings equal global investments.

The projections presented here only reflect the impact of existing policies. Policies that are currently under discussion, such as those included in circular economy roadmaps and resource efficiency frameworks, do not appear in the baseline scenarios presented in this report.[12] Thus, any policy that is not yet implemented or still requires an effort is excluded from the baseline scenarios.

For the energy sector, including fossil fuel use, the projections in this report are aligned with the Current Policies Scenario of the International Energy Agency's World Energy Outlook (IEA, 2017[26]), see Box 2.4. They differ slightly in the long run due to differences in the underlying macroeconomic projections between the two institutions. The various scenarios from the World Energy Outlook illustrate how different assumptions about policies can affect materials projections, for instance including the Nationally Determined Contributions (NDC) of the Paris Agreement on climate change and other planned policies would have many complex implications.

This approach to constructing baseline scenario projections is the appropriate reference point for investigating the costs of inaction and the benefits of policy action. It does not reflect a view on the feasibility and state of recently announced or planned policies. Baseline scenarios implicitly include other government policies that are reflected in the expected trends for the key variables. These underlying policy assumptions drive the modelling assumptions, for instance on the evolution of recycling rates for various materials in different countries, and reflect a combination of the influence of policies and socioeconomic trends. A baseline scenario thus provides a benchmark against which policy scenarios aimed at promoting resource efficiency and the transition to a circular economy can be assessed. A baseline scenario that reflects a continuation of current socioeconomic developments can also be labelled a "business-as-usual" scenario.

Box 2.4. The IEA's World Energy Outlook scenarios

The central baseline scenario energy projections presented in this report rely on the IEA's Current Policies Scenario (CPS). The CPS considers *"only those policies and measures enacted into legislation by mid-2017"* (IEA, 2017[26]) and is thus in line with the definition of a baseline scenario as used in this report.

The IEA also presents a New Policies Scenario (NPS), which includes *"announced policy intentions"*. For example, the greenhouse gas emission mitigation efforts to achieve the Nationally Determined Contributions (NDC) made for the Paris Agreement are included in the NPS, but not in the CPS. The IEA also considers a Sustainable Development Scenario (SDS), which aims at "achieving the main energy-related components of the 2030 Agenda for Sustainable Development". This scenario includes *"urgent action to combat climate change"* and contains additional measures to keep the global temperature increase below 2 degrees Celsius. The IEA does not label any of their scenarios as 'business-as-usual', 'central projection', or 'baseline', arguing that "there is no single story about the future of global energy" (IEA, 2017[26]).

No scenario projection is thus completely free of policies. The decisions about which policies to include or exclude therefore have consequences for the materials use projections and their environmental consequences. The IEA scenarios illustrate this for fossil fuel use and the related consequences for climate change. The influences of the NDCs and additional climate policies on the projections of materials other than fossil fuels are complex, and rely on a detailed understanding of the technology mix underlying climate change mitigation action, and the materials requirements for each of these technologies. This is beyond the scope of the current report.

Figure 2.4. Fossil fuel use and CO2 emissions are projected to increase in the IEA CPS scenario

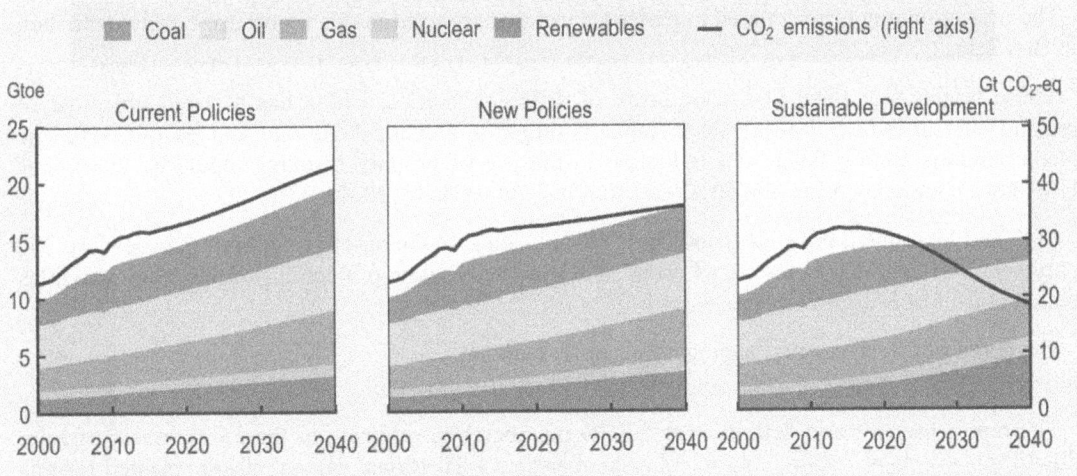

Note: Fuel consumption as reported in the graph does not directly match fossil fuel use as projected in this report and is reported in different units (energy content in Gtoe vs. weight in Gt), but is consistent with these projections. CO_2 emissions are presented in Gt CO_2-equivalent.
Source: (IEA, 2017[26]).

Figure 2.4 clearly shows that the intended policies that are included in the NPS – such as those under the NDCs, as well as air pollution control measures in China and other countries – are projected to reduce the growth of total fuel consumption and CO_2 emissions but not

> achieve an absolute decline. The key difference is a reduction in coal use in Asia (by more than 20%), which has significant positive environmental consequences. The graph also shows that the SDS leads to significant reductions in fossil fuel consumption, and declining CO_2 emissions. Most relevant for this report, under the CPS, all fossil energies (oil, gas and coal) are projected to keep increasing to 2040.

A good dose of humility is warranted when making any model-based projections, especially for long-term projections decades into the future. Many mechanisms that drive long-term economic growth and materials use are not well understood. This means that there are uncertainties about the data input, about long-term projections of the economic drivers, and about the dynamic relationships between these economic drivers and materials use assumed in the projections. There are also likely to be shocks, such as profound or prolonged economic crises or natural disasters, which cannot be foreseen or included in these long-term projections.

This report attempts to embrace uncertainties, as they can help to shed further light on the interactions between economic activity and physical materials use, to highlight which assumptions are most crucial for the projections, and to identify where further work is needed to improve the knowledge base. To deal with uncertainty, this report thus presents a central baseline scenario, complemented with alternative baseline scenarios differing in assumptions on future socioeconomic developments that are presented in dedicated sections of the report.

Notes

[1] The most common terms related to materials use that are used in this report are briefly described in Box 2.1.

[2] These targets are stated in a wide array of different metrics. China has a stated objective of reusing 72% of industrial solid waste. Japan is targeting a cyclical use rate of 17% by 2020. The Netherlands is aiming for a 50% reduction in the use of primary resource inputs by 2030. The USA have a national target of a 50% reduction in food waste by 2030.

[3] GDP and investment are endogenous in ENV-Linkages. In the baseline calibration procedure, the parameters of ENV-Linkages are adjusted such that the baseline projections of these key variables match those of ENV-Growth.

[4] The detail of GTAP sectoral aggregation can be found at https://www.gtap.agecon.purdue.edu/databases/v9/v9_sectors.asp.

[5] The report uses a static definition of OECD membership, referring to the 35 members of June 2018 (http://www.oecd.org/about/membersandpartners/). Historical data are also presented for this static definition and not corrected for historical changes in membership.

[6] An advanced draft of the database was specifically prepared for this project by the experts involved in the database at Vienna University of Economics and Business (WU); the database was publicly released by UNEP in 2018 and is available at www.resourcepanel.org/global-material-flows-database.

[7] Similar problems occur in other databases, such as the OECD's Structural Analysis database (STAN, http://www.oecd.org/sti/ind/stanstructuralanalysisdatabase.htm) and the World Input–Output Database (WIOD, http://www.wiod.org/home).

[8] The data from the Exiobase 3 database (Stadler et al., 2018[16]) are used as input to the Splitcom tool (Horridge, 2005[30]) to disaggregate sectors in the GTAP database (Aguiar, Narayanan and McDougall, 2016[31]) on which the ENV-Linkages relies.

[9] The projections for these detailed sector are based on careful considerations of their supply and demand structure, and the evolution of these structures vis-à-vis evolutions in other sectors. But the model cannot predict whether rapid changes in technologies will occur that will either create new sources of demand or decrease existing sources of demand for specific metals, beyond the technologies explicitly captured in the modelling.

[10] The Exiobase data only contains one casting sector (common to ferrous and non-ferrous metals). So, to be able to split it from the relevant iron & steel and non-ferrous metals sectors in GTAP, the Exiobase sector is split in two, using the relative size of these two sectors (and keeping the exact same production and demand structure).

[11] Section 4.4 in Chapter 4 illustrates how the central baseline scenario differs substantially from a simple linear extrapolation of current production trends for the case of construction in China.

[12] For instance, the EU action plan for the Circular Economy includes revised legislative proposals on waste management that establish a long-term path for waste management and recycling. While several proposals have been adopted (e.g. the Proposed Directive on Packaging Waste or on Landfill), the implementation of the policies is still a long way off. First the actual (revision of) directives need to be adopted, stating concrete targets and timelines. Only then will each Member State implement the regulations and policies needed to reach the targets indicated in the directives.

References

Aguiar, A., B. Narayanan and R. McDougall (2016), "An Overview of the GTAP 9 Data Base", *Journal of Global Economic Analysis*, Vol. 1/1, pp. 181-208, http://dx.doi.org/10.21642/JGEA.010103AF. [31]

Burniaux, J. et al. (1992), "The Costs of Reducing CO2 Emissions: Evidence from GREEN", *Economics Department Working Paper*, No. 115, OECD, Paris, http://www.oecd.org/officialdocuments/publicdisplaydocumentpdf/?cote=OCDE/GD(92)117&docLanguage=En (accessed on 15 January 2018). [27]

Cambridge Econometrics (2014), *Study on modelling of the economic and environmental impacts of raw material consumption*, European Commission Technical Report, http://dx.doi.org/10.2779/74169. [9]

Chateau, J., R. Dellink and E. Lanzi (2014), "An Overview of the OECD ENV-Linkages Model: Version 3", *OECD Environment Working Papers*, No. 65, OECD Publishing, Paris, http://dx.doi.org/10.1787/5jz2qck2b2vd-en. [14]

Chateau, J., C. Rebolledo and R. Dellink (2011), "An Economic Projection to 2050: The OECD "ENV-Linkages" Model Baseline", *OECD Environment Working Papers*, No. 41, OECD Publishing, Paris, http://dx.doi.org/10.1787/5kg0ndkjvfhf-en. [28]

Coulomb, R. et al. (2015), "Critical minerals today and in 2030: an analysis for OECD countries", *Environment Working Paper No. 91*, http://www.oecd.org/environment/workingpapers.htm (accessed on 10 January 2018). [6]

Crowson, P. (2011), "Mineral reserves and future minerals availability", *Mineral Economics*, Vol. 24/1, pp. 1-6, http://dx.doi.org/10.1007/s13563-011-0002-9. [22]

Distelkamp, M. and M. Meyer (2016), "Quantitative assessment of pathways to a resource-efficient and low-carbon Europe", *GWS Discussion Papers*, No. 2016/10, https://www.econstor.eu/handle/10419/156301 (accessed on 09 February 2018). [11]

Giljum, S. (2014), "Global patterns of material flows and their socio-economic and environmental implications: a MFA study on all countries world-wide from 1980 to 2009", *Resources*, Vol. 3/1, pp. 319-339, http://dx.doi.org/10.3390/resources3010319. [20]

Guillemette, Y. and D. Turner (2018), "The Long View: Scenarios for the World Economy to 2060", *OECD Economic Policy Papers*, No. 22, OECD Publishing, Paris, http://dx.doi.org/10.1787/b4f4e03e-en. [15]

Horridge, M. (2005), *SplitCom: Programs to disaggregate a GTAP sector*. [30]

Hu, J., S. Moghayer and F. Reynès (2015), *Report about integrated scenario interpretation EXIOMOD/LPJmL results*, POLFREE Deliverable D3.7B., http://www.polfree.eu/publications/publications-2014/report-d37b. [10]

Humphreys, D. (2011), "Emerging miners and their growing competitiveness", *Mineral Economics*, Vol. 24/1, pp. 7-14, http://dx.doi.org/10.1007/s13563-011-0005-6. [24]

Hyman, R. et al. (2003), "Modeling non-CO2 Greenhouse Gas Abatement", *Environmental Modeling and Assessment*, Vol. 8/3, pp. 175-186, http://dx.doi.org/10.1023/A:1025576926029. [29]

IEA (2017), *World Energy Outlook 2017*, OECD Publishing, Paris/IEA, Paris, http://dx.doi.org/10.1787/weo-2017-en. [26]

Kelly, T. (2014), *Historical statistics for mineral and material commodities in the United States (2016 version): U.S. Geological Survey Data Series 140*, https://minerals.usgs.gov/minerals/pubs/historical-statistics/ (accessed 22 May 2018). [25]

Krausmann, F. et al. (2009), "Growth in global materials use, GDP and population during the 20th century", *Ecological Economics*, Vol. 68, pp. 2696-2705, http://dx.doi.org/10.1016/j.ecolecon.2009.05.007. [2]

Mannaerts, H. (2004), *Environmental policy analysis with STREAM: a partial equilibrium model for material flows in the economy*, MIT Press. [17]

McCarthy, A., R. Dellink and R. Bibas (2018), *The Macroeconomics of the Circular Economy Transition: A Critical Review of Modelling Approaches*, OECD Environment Working Papers, http://dx.doi.org/10.1787/af983f9a-en. [13]

Meyer, B., M. Distelkamp and T. Beringer (2015), "Report about integrated scenario interpretation GINFORS / LPJmL results", POLFREE Deliverable D3.7A., http://www.polfree.eu. [18]

OECD (2016), *Policy Guidance on Resource Efficiency*, OECD Publishing, Paris, http://dx.doi.org/10.1787/9789264257344-en. [3]

OECD (2015), *Material Resources, Productivity and the Environment*, OECD Green Growth Studies, OECD Publishing, Paris, http://dx.doi.org/10.1787/9789264190504-en. [1]

OECD (2012), *OECD Environmental Outlook to 2050: The Consequences of Inaction*, OECD Publishing, Paris, http://dx.doi.org/10.1787/9789264122246-en. [5]

OECD (2008), *Measuring material flows and resource productivity; Volume I, The OECD Guide*, OECD, PARIS, https://www.oecd.org/environment/indicators-modelling-outlooks/MFA-Guide.pdf (accessed on 06 February 2018). [7]

Schandl, H. (2017), "Global Material Flows and Resource Productivity: Forty Years of Evidence", *Journal of Industrial Ecology*, Vol. 46/1, http://dx.doi.org/10.1111/jiec.12626. [21]

Stadler, K. et al. (2018), "EXIOBASE 3: Developing a Time Series of Detailed Environmentally Extended Multi-Regional Input-Output Tables", *Journal of Industrial Ecology*, http://dx.doi.org/10.1111/jiec.12715. [16]

UN (2017), "World Population Prospects: key findings and advance tables", https://esa.un.org/unpd/wpp/publications/Files/WPP2017_KeyFindings.pdf (accessed on 18 May 2018). [4]

UNEP (2017), *Resource Efficiency: Potential and Economic Implications. A report of the International Resource Panel.*, Ekins, P., Hughes, N., et al.. [12]

USGS (2017), *Critical mineral resources of the United States—Economic and environmental geology and prospects for future supply*, http://dx.doi.org/10.3133/pp1802. [23]

Wiedmann, T. et al. (2015), "The material footprint of nations.", *Proceedings of the National Academy of Sciences of the United States of America*, Vol. 112/20, pp. 6271-6, http://dx.doi.org/10.1073/pnas.1220362110. [19]

Winning, M. et al. (2017), "Towards a circular economy: insights based on the development of the global ENGAGE-materials model and evidence for the iron and steel industry", *International Economics and Economic Policy*, Vol. 14/3, pp. 383-407, http://dx.doi.org/10.1007/s10368-017-0385-3. [8]

Annex 2.A. The modelling framework

2.A.1 The modelling approach to produce economic and materials projections

The process to produce economic and materials projection relies on a suite of modelling tools, as described in Section 2.2. As a first step, the OECD ENV-Growth model, a macroeconomic growth model based on a conditional convergence framework, projects global macroeconomic trends by country. It does so by using a set of assumptions on long-term macro trends such as demography or technical progress, in addition to short-term forecasts and long-term projections of the OECD Economics Department and the International Monetary Fund (IMF).

The output of ENV-Growth is an essential input for the OECD's multisectoral ENV-Linkages model, which focuses on the evolution of sectoral and regional economic activity and the associated economy-environment interactions in the coming decades. In fact, as a second step, the projections of GDP, investment and trade balances from ENV-Growth, are imposed on ENV-Linkages in the calibration of the baseline scenarios. These GDP projections are then endogenised in ENV-Linkages to allow assessment of the impacts of alternative scenarios. Thus, the exogenous parameters in ENV-Linkages, e.g. on productivity, are adjusted in the calibration process to mimic the macroeconomic growth paths from ENV-Growth. ENV-Linkages and ENV-Growth are thus mutually consistent. As a last step, as ENV-Linkages is enhanced to include data on materials use and recycling. In ENV-Linkages material flow projections follow the evolution of the production and consumption of goods in different sectors and regions.

2.A.2 The ENV-Growth model

The OECD regularly produces country-specific long-term macroeconomic projections for OECD countries using the Long Term model (Guillemette and Turner, 2018[15]). These provide a reference scenario for socioeconomic development against which the consequences of alternative policy settings can be projected. The OECD's ENV-Growth model extends the methodology used so far to produce economic growth projections for both OECD and non-OECD countries. The model explicitly considers energy and some natural resources (oil and gas) as productive inputs. The methodology provides long-term pathways of national income until 2060 for 230 individual countries.

The ENV-Growth modelling framework for projecting future global and country-specific GDP and per-capita income levels is based on a neoclassical model of exogenous growth augmented with accumulating human capital (i.e. augmented Solow growth model). Countries' income levels converge towards their country-specific long-term frontier ("conditional convergence"), which is determined by the key drivers of economic growth. Short- to medium-term (2012-2017) economic projections are in line with the OECD and the International Monetary Fund (IMF) forecasts as of Summer 2018 and Spring 2018, respectively. The projections typically show higher growth rates for developing countries and partial income convergence across countries over the century (Figure 2.A.1; Figure 2.A.2).

Figure 2.A.1. Stylised representation of conditional income convergence

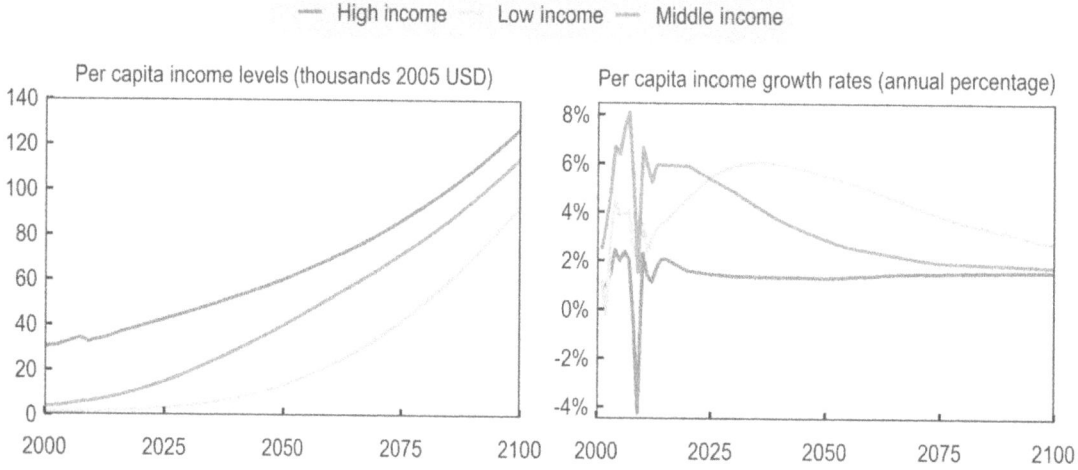

Source: OECD ENV-Growth model.

Figure 2.A.2. Global distribution of per-capita income

Thousands USD in 2005 exchange rates

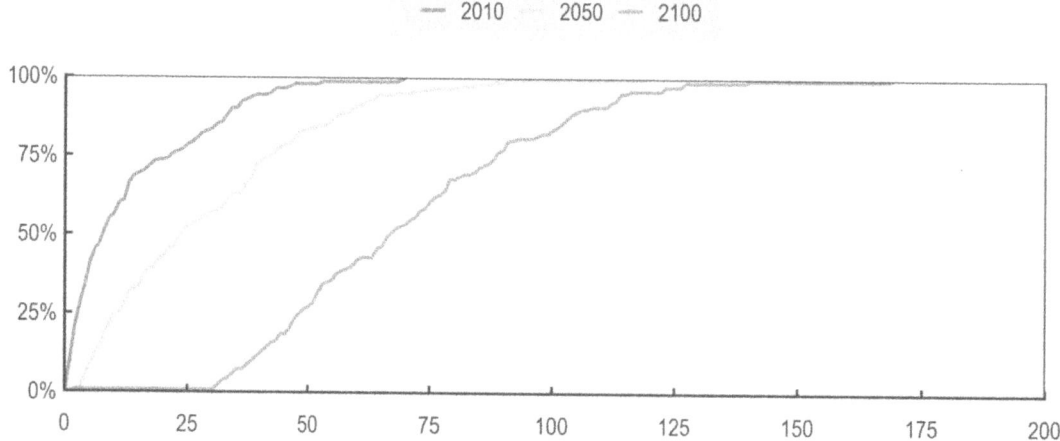

Source: OECD ENV-Growth model.

The OECD's ENV-Growth model considers six drivers of long-term economic growth (depicted in Figure 2.A.3): Total factor productivity (TFP), as an indicator of exogenous technical progress; Physical capital, as driven by standard capital accumulation; Labour, as driven by human capital, which depends on education, and employment, which depends on demographic trends, labour participation rates and unemployment scenarios; Energy demand, as driven by autonomous energy efficiency; Natural resource revenues stemming from extraction and processing of oil and gas.

Figure 2.A.3. The ENV-Growth Modelling framework overview

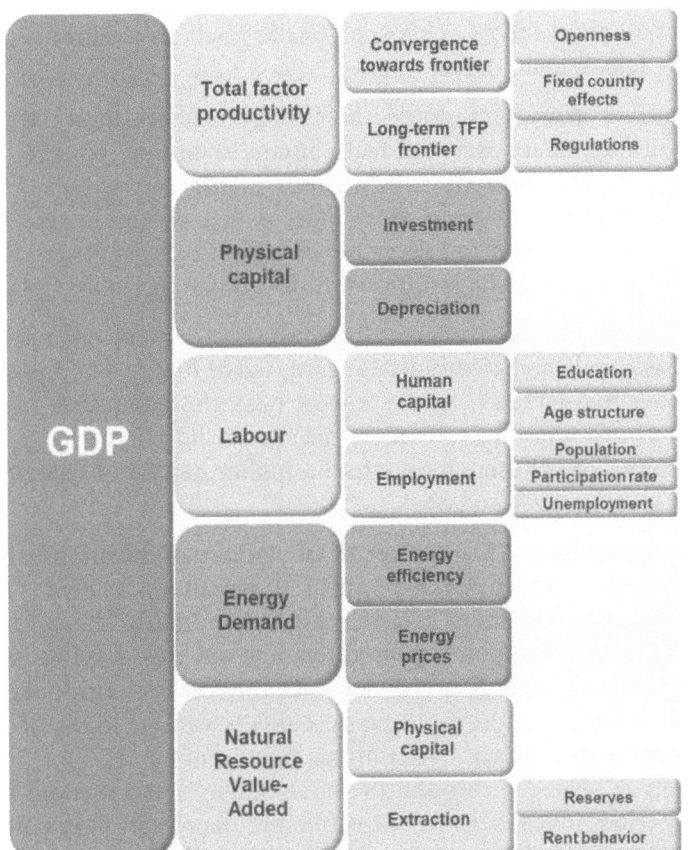

2.A.3 The ENV-Linkages model

The OECD's in-house dynamic CGE model - ENV-Linkages - is used as the basis for the assessment of the economic consequences of environmental impacts until 2060 as well as to study the economic consequences of environmental policies.

ENV-Linkages is a multi-sectoral, multi-regional model that links economic activities to energy and environmental issues. It is the successor to the OECD GREEN model for environmental studies (Burniaux et al., 1992[27]). A more comprehensive model description is given in Chateau, Dellink and Lanzi (2014[14]); whereas a description of the baseline scenario construction procedure is given in Chateau, Rebolledo and Dellink (2011[28]).

Production in ENV-Linkages is assumed to operate under cost minimisation with perfect markets and constant return to scale technology. The production technology is specified as nested Constant Elasticity of Substitution (CES) production functions in a branching hierarchy. This structure is replicated for each output, while the parameterisation of the CES functions may differ across sectors. The nesting of the production function for the agricultural sectors is further re-arranged to reflect substitution between intensification (e.g. more fertiliser use) and extensification (more land use) of crop production; or between intensive and extensive livestock production. The structure of electricity production assumes that a representative electricity producer maximizes its profit by using the different available technologies to generate electricity using a CES specification with a large degree of substitution. The structure of non-fossil electricity technologies is similar to that of other sectors, except for a top nest combining a sector-specific resource with a sub-nest of all other inputs. This specification acts as a capacity constraint on the supply of the electricity technologies.

The model adopts a putty/semi-putty technology specification, where substitution possibilities among factors are assumed to be higher with new vintage capital than with old vintage capital. In the short run this ensures inertia in the economic system, with limited possibilities to substitute away from more expensive inputs, but in the longer run this implies relatively smooth adjustment of quantities to price changes. Capital accumulation is modelled as in the traditional Solow/Swan neo-classical growth model.

The energy bundle is of particular interest for analysis of climate change issues. Energy is a composite of fossil fuels and electricity. In turn, fossil fuel is a composite of coal and a bundle of the "other fossil fuels". At the lowest nest, the composite "other fossil fuels" commodity consists of crude oil, refined oil products and natural gas. The value of the substitution elasticities are chosen as to imply a higher degree of substitution among the other fuels than with electricity and coal.

Household consumption demand is the result of static maximization behaviour which is formally implemented as an "Extended Linear Expenditure System". A representative consumer in each region– who takes prices as given– optimally allocates disposal income among the full set of consumption commodities and savings. Saving is considered as a standard good in the utility function and does not rely on forward-looking behaviour by the consumer. The government in each region collects various kinds of taxes in order to finance government expenditures. Assuming fixed public savings (or deficits), the government budget is balanced through the adjustment of the income tax on consumer income. In each period, investment net-of-economic depreciation is equal to the sum of government savings, consumer savings and net capital flows from abroad.

International trade is based on a set of regional bilateral flows. The model adopts the Armington specification, assuming that domestic and imported products are not perfectly substitutable. Moreover, total imports are also imperfectly substitutable between regions of origin. Allocation of trade between partners then responds to relative prices at the equilibrium.

Market goods equilibria imply that, on the one side, the total production of any good or service is equal to the demand addressed to domestic producers plus exports; and, on the other side, the total demand is allocated between the demands (both final and intermediary) by domestic producers and the import demand.

CO_2 emissions from combustion of energy are directly linked to the use of different fuels in production. Other GHG emissions are linked to output in a way similar to Hyman et al. (2003[29]). The following non-CO_2 emission sources are considered: (i) methane from rice cultivation, livestock production (enteric fermentation and manure management), fugitive methane emissions from coal mining, crude oil extraction, natural gas and services (landfills and water sewage); (ii) nitrous oxide from crops (nitrogenous fertilisers), livestock (manure management), chemicals (non-combustion industrial processes) and services (landfills); (iii) industrial gases (SF_6, PFCs and HFCs) from chemicals industry (foams, adipic acid, solvents), aluminium, magnesium and semi-conductors production

ENV-Linkages is fully homogeneous in prices and only relative prices matter. All prices are expressed relative to the numéraire of the price system that is arbitrarily chosen as the index of OECD manufacturing exports prices. Each region runs a current account balance, which is fixed in terms of the numéraire. One important implication from this assumption in the context of this paper is that real exchange rates immediately adjust to restore current account balance when countries start exporting/importing emission permits.

As ENV-Linkages is recursive-dynamic and does not incorporate forward-looking behaviour, price-induced changes in innovation patterns are not represented in the model. The model does, however, entail technological progress through an annual adjustment of the various productivity parameters in the model, including e.g. autonomous energy efficiency and labour productivity improvements. Furthermore, as production with new capital has a relatively large degree of flexibility in choice of inputs, existing technologies can diffuse to other firms. Thus, within the CGE framework, firms choose the least-cost combination of inputs, given the existing state of technology. The capital vintage structure also ensures that such flexibilities are large in the long run than in the short run.

The sectoral aggregation of the model adopted in this report is given in Table 2.A.1.

Table 2.A.1. Sectoral aggregation of ENV-Linkages

Agriculture, Fisheries and Forestry	Manufacturing
Paddy Rice	Food Products
Wheat and Meslin	Textiles
Other Grains	Wood products
Vegetables and Fruits	Chemicals
Oil Seeds	Pulp, Paper and Publishing products
Sugar Cane and Sugar Beet	Non-metallic Minerals
Fibres Plant	Fabricated Metal products
Other Crops	Electronics
Cattle and Raw Milk	Motor Vehicles
Other Animal products	Other Transport Equipment
Fisheries	Other Machinery and Equipment
Forestry	Recycling
Non-manufacturing Industries	Iron and Steel - Primary
Coal extraction	Iron and Steel – Secondary
Crude Oil extraction	Aluminium – Primary
Natural Gas extraction	Aluminium – Secondary
Other Mining	Copper – Primary
Petroleum and Coal products	Copper – Secondary
Gas distribution	Other Non-ferrous Metals – Primary
Water Collection and Distribution	Other Non-Ferrous metals – Secondary
Construction	Other Manufacturing
Electricity Transmission and Distribution	**Services**
Electricity Generation (8 technologies)	Land Transport
Electricity generation: Nuclear Electricity; Hydro (and Geothermal); Solar; Wind; Coal-powered electricity; Gas-powered electricity; Oil-powered electricity; Other (combustible renewable, waste, etc).	Air Transport
	Water Transport
	Business Services
	Other Services (incl. Government)

Chapter 3.

Projecting the economic baseline scenario

This chapter outlines the main economic mechanisms that drive the economic projections in the baseline scenario. It begins by presenting the socioeconomic trends at the macroeconomic level. These trends include sociodemographic assumptions (population and labour force participation projections) and macroeconomic assumptions (capital accumulation and technical progress). Next, the changes in demand patterns that affect the sectoral composition of the economy are discussed, with an emphasis on the projected shift of demand towards services in the coming decades. The third section discusses the evolution of production processes, which shows changes in the composition of the inputs to production as a result of technical progress and the transformation of input relative prices. Taken together, changes in demand patterns and production processes explain the structural change of the economy over time. The final section outlines the uncertainty surrounding the socioeconomic drivers.

This document, as well as any data and any map included herein, are without prejudice to the status of or sovereignty over any territory, to the delimitation of international frontiers and boundaries and to the name of any territory, city or area.

KEY MESSAGES

The economic projections presented in this chapter form a base from which to model the evolution of future materials use. There are three main mechanisms that influence the economic projections in the baseline scenarios: socioeconomic evolution, changes in demand patterns, and changes in production processes.

Projections and trends

- Over the coming decades, economic growth projections will be driven by changes in demographic projections as well productivity growth. These projected trends also imply economic convergence across countries (where – conditional on local circumstances – countries gradually catch up to the most developed countries in terms of living standards). The figure below illustrates that almost all countries in the world are projected to significantly increase their income levels in 2060 relative to 2010, with emerging economies growing the most. India and the rest of Southeast Asia are gradually taking over from China as the engines of global growth.

Figure 3.1. GDP is projected to increase significantly in all countries but global growth is slowing down

GDP trillion USD (2011 PPP)

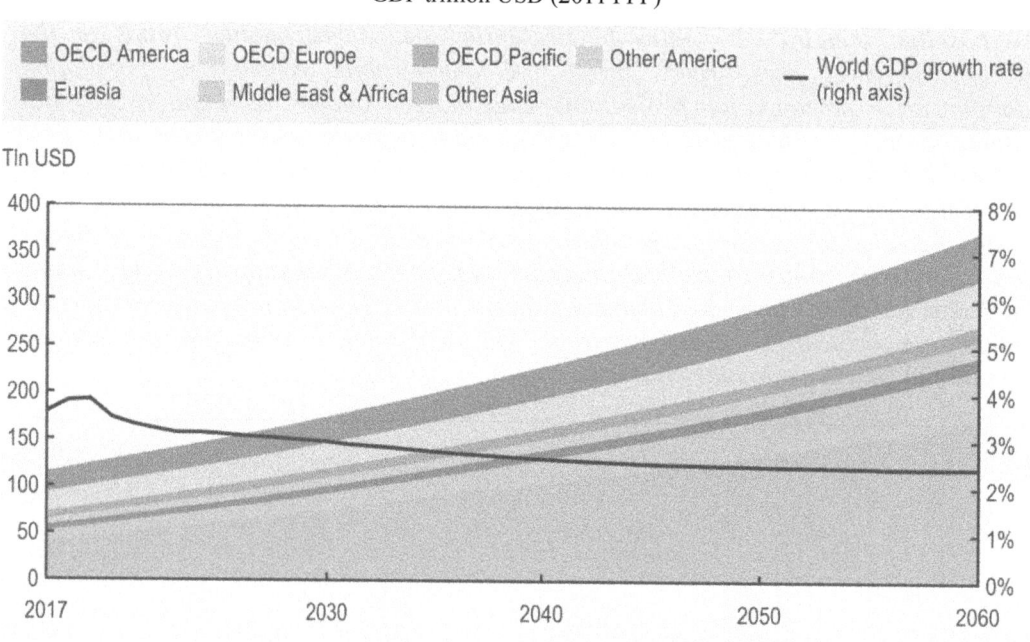

Source: OECD ENV-Linkages model.

StatLink https://doi.org/10.1787/888933884498

- Economic projections are also characterized by changes in the basket of commodities demanded by households and firms. The gradual shift of global demand from manufacturing and agricultural goods towards services is projected to continue. The main explanation is that the share of demand for services in total expenditures increases with the level of income. This also explains the stronger shift of demand in non-OECD countries where the share of demand for services in 2017 is lower than in OECD, and the projected growth is higher. The increasing share of demand for services is also a consequence of population ageing, mainly in European countries, Japan and China. The baseline economic projections also show a growing share of service expenses in total production cost in all economies, reflecting servitisation of manufacturing production and growing digitalization.

- Finally, the economic projections show modifications in production processes (in the mix of inputs used in production or in the shift between technologies). These changes in the production of all goods and services will be the result of firms' adaptation to future changes in production factor costs and innovations. One key evolution is the increasing share of labour cost in total production cost of manufacturing goods, another is the decrease in average cost of producing manufacturing goods relative to the cost of other goods and services. These trends are stronger in non-OECD countries, where a higher rate of convergence also leads to more marked changes in relative productivity of the different input of production over time.

Overall, the projections indicate a regional shift of GDP: the share of OECD countries in global GDP is projected to fall to 31% in 2060, from a 44% share in 2017 (which was 61% in 2000), while in the same period the share of non-OECD South-East Asia is projected to increase from 35% to 46%.

Areas of uncertainty

Two key sources of uncertainties in the projections are explored:

1. Alternative demographic projections, based on low and high population scenarios from the UN population prospects.

2. Slower or faster convergence of labour efficiency (and hence income) across countries.

Modelling these alternative assumptions leads to world GDP variations from the central baseline scenario of -3% to 3% in 2030 and of -19% to 18% in 2060.

Policy implications

The analysis of the uncertainties surrounding the long run macroeconomic growth projections enables a better understanding of the robustness of the projections of economic activity and materials use. Furthermore, this analysis helps to place the potential impact of policies on economic performance in context: policy impacts on GDP may or may not be relatively small when compared to the uncertainty surrounding long-term growth. This does not imply that policy costs are irrelevant. Rather, the expected net benefits of policies should be maximised, e.g. through implementing policies that work well in any socioeconomic scenario.

The projections are also a suitable reference point for the economic assessment of policies to reduce materials use and its environmental consequences.

3.1. Population and economic growth will see major regional shifts

When looking at the future evolution of income and economic activities, a range of socioeconomic trends need to be taken into account. These trends include sociodemographic transformations and evolutions, including population growth. Another key trend is the gradual convergence of living standards (per-capita income levels) towards those of the most advanced economies, conditional on country-specific conditions (hereafter labelled income convergence). The growth of population and per-capita income together determine the scale of economic activity and thus the scale of materials use. These future trends in socioeconomic variables are the basis for the central baseline scenario presented in this report.[1]

World population has been increasing in recent decades and is projected to continue increasing in the coming decades. The central baseline scenario projects global population will reach more than 10 billion people by 2060 (see Figure 3.2), drawing on the "medium scenario" of the World Population Prospects (UN, 2017[1]) and the central scenario of Eurostat projections for European countries (Eurostat, 2018[2]).[2] The pace of population growth is slowing between 2017 and 2060, which contrasts with the past 40 years of strong growth. Over the next decades (between 2017 and 2060), global population is projected to grow by 0.7% per year on average, while the growth rate was 1.4% per year during the period 1980-2017.

Figure 3.2. World population is projected to keep growing but less rapidly than in the past

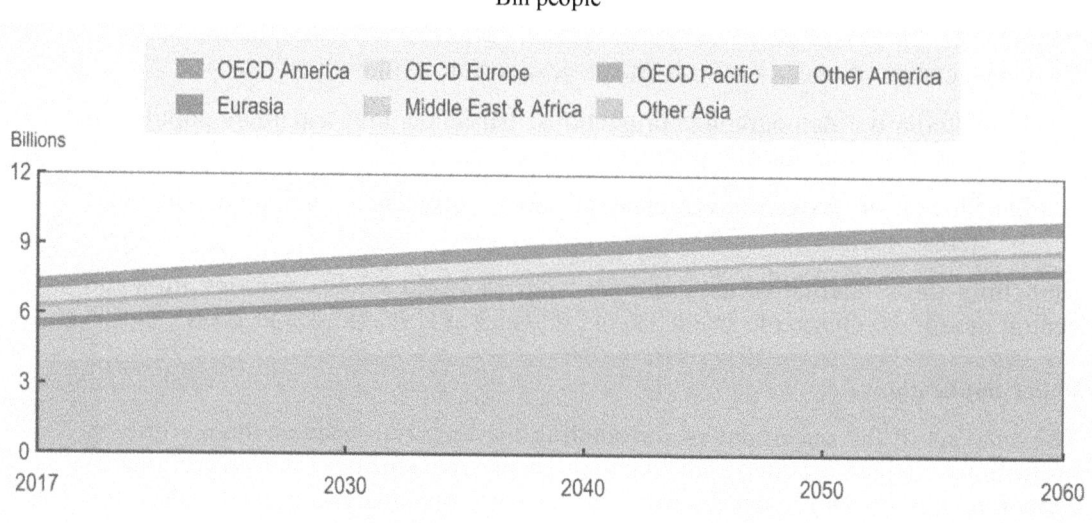

Source: Own calculation from The World Population Prospects: 2017 Revision (UN, 2017[1]) and Eurostat (Eurostat, 2018[2]).

StatLink https://doi.org/10.1787/888933884517

This decline in population growth applies to all countries. However, population growth trends will vary across countries. Some countries with the most advanced demographic transition are projected to even face negative growth (several European countries, Japan, Korea, and China). At the other extreme, Sub-Saharan Africa (grouped with the other parts of Africa and the Middle East in the figure) is projected to experience very high population growth (over 2% per year over 2017-2060). As a result, more than 29% of world population in 2060 is projected to be settled in Africa, compared to 17% in 2017. In contrast, the OECD share shrinks from 17% in 2017 to 14% in 2060 (see Table 3.A.1 in Annex 3.A for the detail by region).

In the coming decades, the global population is projected to not only increase but also to become wealthier. The macroeconomic projections for OECD and G20 countries match the long-term macroeconomic projections of the OECD Economics Department (Guillemette and Turner, 2018[3]). For the remaining countries, projections are provided by the ENV-Growth model. Living standards (measured as GDP per capita) are projected to increase over the entire period, with most countries gradually converging towards 2017 OECD levels (Figure 3.3). The improvements in living standards over the 2017-2060 projection period (blue bars) are projected to be greater for countries that currently have lower levels of per-capita GDP (those to the right of the graph, since the figure is sorted by GDP per capita in 2017 in yellow). The poorer countries at the beginning of the period are thus projected to show important gains in living standards (including Sub-Saharan African countries[3], India, and other non-OECD Asian countries). Global income per capita is projected to almost reach the 2017 OECD level of living standards by 2060.

Figure 3.3. Living standards are projected to gradually converge

Real GDP per capita in USD (2011 PPP), sorted by GDP per capita in 2017

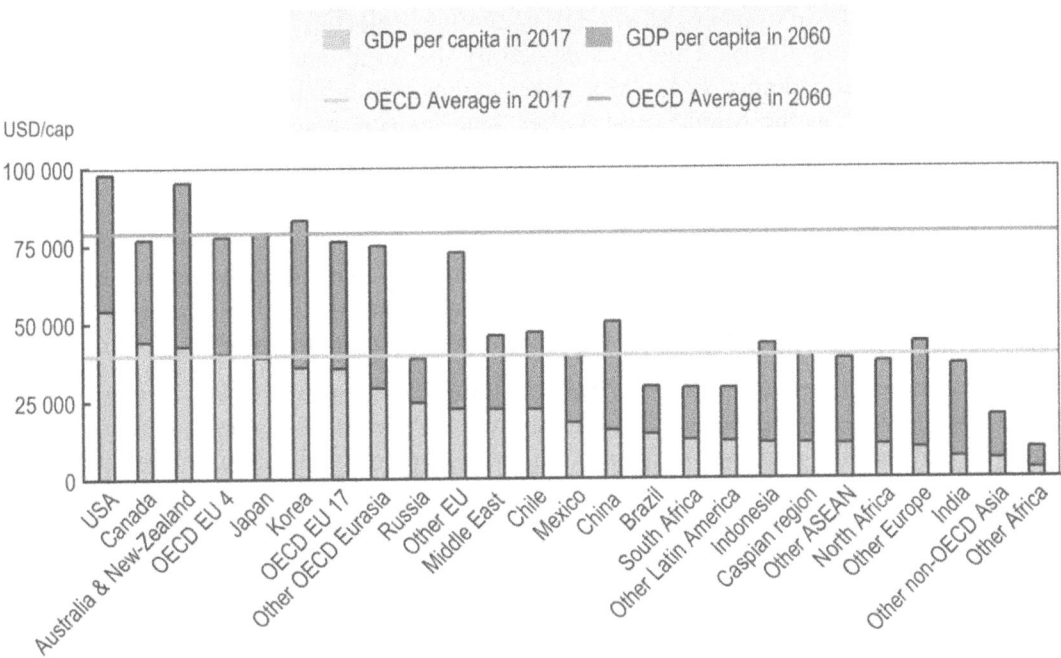

Note: See Table 2.1 for regional definitions. In particular, OECD EU 4 includes France, Germany, Italy and the United Kingdom. OECD EU 17 includes the other 17 OECD EU member states. Other OECD Eurasia includes the EFTA countries as well as Israel and Turkey. Other EU includes EU member states that are not OECD members. Other Europe includes non-OECD, non-EU European countries excluding Russia. Other Africa includes all of Sub-Saharan Africa excluding South Africa; in the text, the term Other Africa is replaced with Sub-Saharan Africa to improve readability. Other non-OECD Asia includes non-OECD Asian countries excluding China, India, ASEAN and Caspian countries.
Source: ENV-Growth model (OECD Environment Directorate) and OECD Economics Department (Guillemette and Turner, 2018[3]).

StatLink https://doi.org/10.1787/888933884536

Two categories of countries deviate from this pattern. Countries that are fossil-fuel exporters are projected to underperform compared to the standard pattern, as fossil fuel revenues do not grow as rapidly as other contributing factors to GDP. Countries in this category include the Russian Federation (hereafter Russia), Brazil and Middle Eastern countries. In contrast, European countries that are currently in a phase of integration to the European Union (EU), especially those labelled as "Other EU"[4], are projected to overperform.

Living standards in developing economies will still be far from those of OECD economies at the end of the time horizon, despite this convergence process. This can be seen in Figure 3.3, which presents real GDP per capita in 2060 by region (shown as stacked bars in 2060, while the OECD average is presented as a horizontal line). Some countries are projected to not reach 2017 OECD levels by 2060; these include countries in Latin America, Other non-OECD Asia, and Sub-Saharan Africa. Mexico, Russia and India, among others, are projected to reach in 2060 a level close to the 2017 OECD living standards.

As a result of increasing population and living standards, global GDP increases, as shown in Panel A of Figure 3.4. GDP increases in all regions, even in countries where population is declining, since the growth of GDP per capita has a larger impact than population changes.

The share of OECD countries in global GDP in 2060 is projected to fall to 31% from 44% in 2017 (from 61% in 2000).[5] This is explained by the large increase in the share of the Asian developing economies, and – to a lesser extent – Sub-Saharan African countries. Other regions, such as the Middle East, Other America (i.e. non-OECD Latin America) and the Eurasia group of countries are not projected to see their share in global GDP increase significantly.

This pattern results from the fact that countries with more dynamic demographic changes, especially faster growing populations, are also countries with high gains in GDP per capita, so their shares in world total GDP increase substantially. It therefore appears that projected trends of GDP per capita and population growth generally move together.

The central baseline scenario projects that the global GDP growth rate will slow down and stabilise just below 2.5% after 2030, as shown in Panel B of Figure 3.4. While India and large parts of Sub-Saharan Africa are projected to record high growth rates and then become important drivers of world growth in the 2020-2040 period, the projected slowdown of the Chinese economy after 2025 dominates. From around 2040, the most dynamic region is projected to be Sub-Saharan Africa, but its increasing share in world GDP growth is not sufficient to counterbalance the slowdown of China's economic growth in this scenario.

Figure 3.4. Emerging economies drive the projected global GDP growth

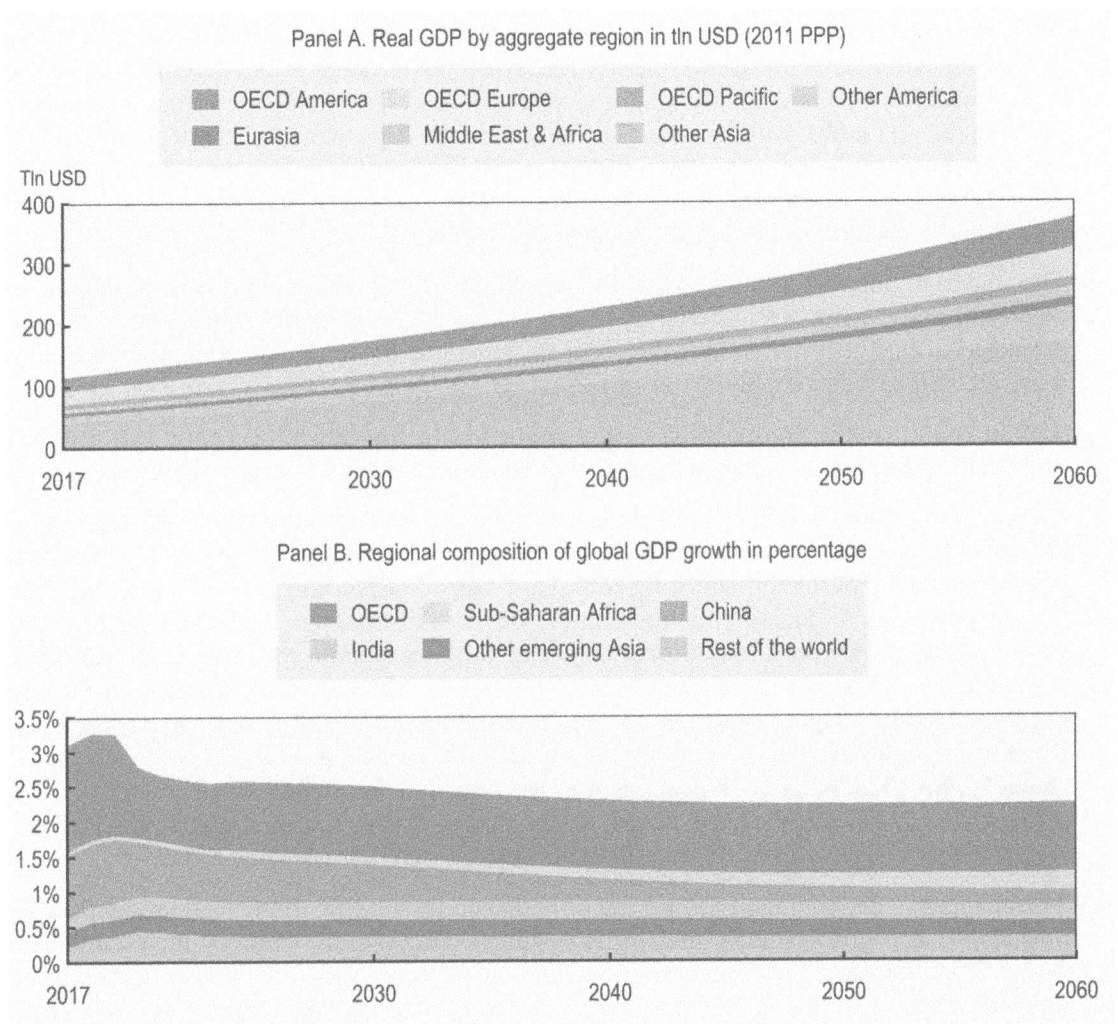

Note: Panel B uses a custom aggregation of regions to highlight the contribution of China and India.
Source: OECD ENV-Linkages model; short-term forecasts by OECD Economics Department (as of Summer 2018) and IMF (as of Spring 2018).

StatLink https://doi.org/10.1787/888933884555

3.2. The services sector will drive demand growth

3.2.1. Rising incomes mean new consumption patterns

An increase in GDP does not mean that the proportion of each good produced and consumed remains constant. The structure of the economy evolves because living standards transform preferences; because society is changing with increasing ageing and urbanisation, and also because the nature of production is evolving, relying more on research and development (R&D) and services[6] expenses. In particular, the model projects an increasing demand for services by households, government and firms. This section focuses on the projected changes in final demand (i.e. demand by households, government and for investment purposes), while the next focuses on changes in intermediate demand by firms.

As income per capita increases, final demand patterns change. The share of necessary commodities (food and agricultural products) in total expenditure decreases, while the share of luxury goods – such as recreational and leisure activities and other services (including health and education) – increases. This conventional effect is reinforced in the central baseline scenario by the assumption that in emerging and developing economies preferences gradually shift towards OECD standards. This includes changes in the size and direction of government expenditures, as well as shifts in household expenditures towards services. These preference shifts are partially driven by income growth, but also reflect the projected further digitalisation of the economy.

The share of manufacturing goods in households' total expenditures is projected to decline slightly, but more importantly, expenditures on durable and equipment goods are projected to change. For example they will shift away from equipment and paper, towards more electronics and vehicles.

Similar trends in the composition of government and investment expenditures are also projected, which include increasing shares of education and R&D expenditures.

Ageing also induces a shift of household and government demand towards more services, not least for health and other long-term elderly care expenditures. Even if public and private spending on health and long-term care vary considerably across countries, they are all projected to increase in the future (de la Maisonneuve and Oliveira Martins, 2014[4]). The projected increase of health and long-term care spending is driven by a combination of ageing and other demographic factors, as well as the increase in income per capita and technical progress (de la Maisonneuve and Oliveira Martins, 2014[4]). Regardless of the drivers, the result is an increase in the demand for the "other services" category, which includes health care as well as education and public services.

3.2.2. The "servitisation" of manufacturing is a significant trend

The changes in demand patterns are not only driven by modifications of final demand by households and governments, and for investment, but also by changes in intermediate demand, i.e. demand for produced goods and services by firms. This is reflected in an intensification of services as inputs to all sectors (including manufacturing processes), known as the "servitisation of manufacturing" (Pilat and Nolan, 2016[5]). Both servitisation of manufacturing and service digitalisation result from the Information and Communication Technology (ICT) revolution, the intensification of R&D expenses, and the growth of the sharing economy.[7]

This intensification of services in the economy goes further: it includes the shift in business models towards more and more services. The business of car companies for instance is increasingly geared towards services such as insurance, credit, and maintenance.

The main consequence of this structural transformation is that the services sectors, and especially the business services sector, are projected to grow faster than the rest of the economy in all countries over the period 2017-2060 (Figure 3.5).

Figure 3.5. Demand for services is projected to increase more than the economy-wide average

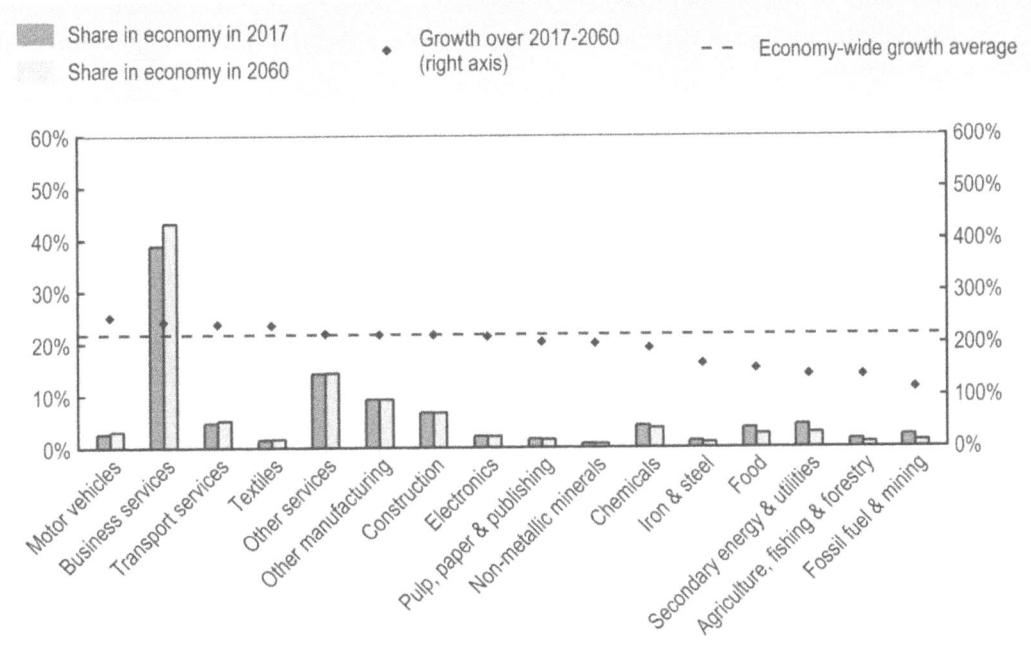

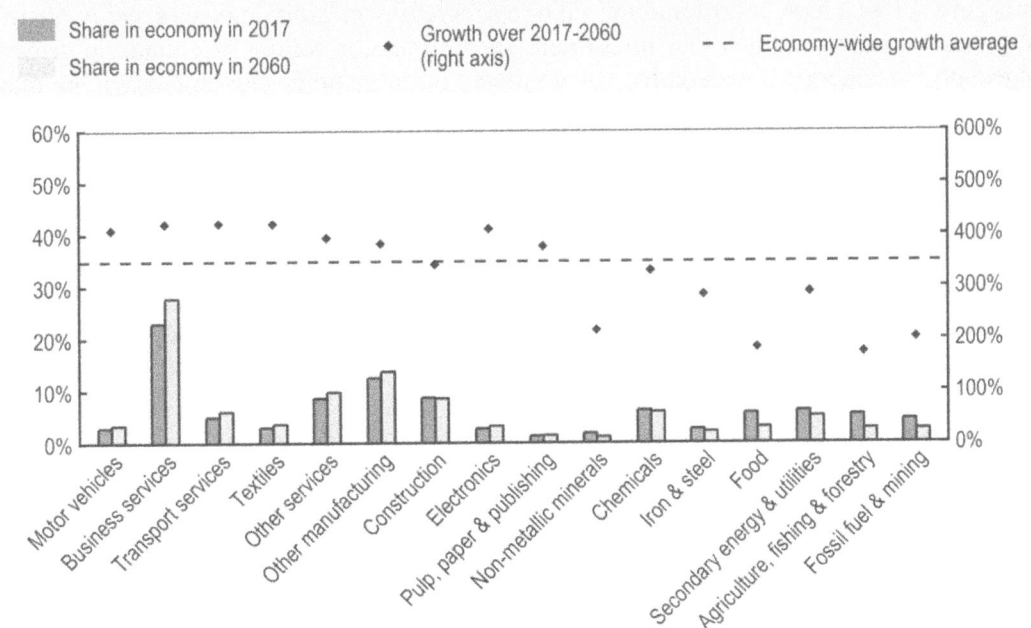

Source: OECD ENV-Linkages model.

StatLink https://doi.org/10.1787/888933884574

In contrast, the output of the fossil fuel and mining sectors, as well as of energy intensive industries[8] is projected to increase less than the economy-wide average, mainly in OECD countries but also in emerging economies. Similarly, the share of food and agricultural goods in total expenditures is projected to diminish significantly. However, the global demand for these goods is still projected to increase by 65% by 2060 compared with 2017 levels: agricultural and food expenditures increase, but less rapidly than expenditures on other goods and services.

3.3. Production processes will rely more on new technologies and services

The GDP changes described in Section 3.1 are largely driven by the evolution of the main primary factors of production (capital and labour) as well as by technical progress. These changes can come from a wide range of drivers, including continued efforts to optimise existing production processes, adopting new business models, and the spreading of best available techniques. The change in GDP per capita can be broken down into changes in employment levels, in labour efficiency and in the amount of capital per worker (Figure 3.7).[9]

Changes in labour efficiency have the strongest influence on per-capita GDP growth. Long run labour efficiency gains are assumed to be driven by country-specific progress in education levels, investment in innovation, and improvement in the quality of institutions and market regulations, as well as other determinants.[10] As shown in Figure 3.7, and in accordance with traditional growth theory,[11] in the long run the gains in living standards (diamond marks) converge (see Annex 3.A).

However, in the short and medium run (2017-2030), the process of catching up through increases in capital-to-output ratios plays a non-negligible role. This mechanism is visible in Figure 3.7 as a high contribution to GDP by increases in capital per worker. A relative shortage of capital implies that investments are the major source of economic growth, especially in emerging economies. In contrast, investment is slowing down in more advanced economies, not only because equipment and infrastructure expenditures have largely already been undertaken, but also due to the reduction of saving rates that characterise these ageing societies (see Box 3.1).

Furthermore, in the short and medium term employment rates fluctuate and influence the dynamics of GDP per capita. In many regions, employment growth makes a positive contribution to growth, but in countries with significant ageing, employment changes become a drag on economic growth, as the share of the working age population in the overall population declines (see details in Annex 3.A).

Box 3.1. The macroeconomic impacts of ageing

Having a higher share of elderly people in the total population can have macroeconomic consequences that can affect materials use. While the direct impact of ageing on the reduction of potential employment growth has consequences for the scale of economic activities, it does not on its own affect the intensity of materials use. However, the impact of ageing on savings does have implications for materials use.

Several elements drive saving rates in the long run: inflation rates, time preferences, risk aversion, and changes in the age structure. These factors are either susceptible to conjuncture or stable over time. The one exception is the ageing, which is both an established characteristic of demographic projections and a determinant of saving rates. Indeed, during the demographic transition, savings rates are projected to steadily decrease. Regions with the fastest-increasing old-age dependency ratios (see Figure 3.A.1 in Annex 3.A) have the fastest-decreasing net saving rates, namely Japan and Korea, the EU, Russia and China (IMF, 2004[6]).

Thus, the share of household consumption in GDP is projected to increase and then stabilize in these countries, as shown in the figure below. This is to a lesser extent also the case for government expenses. In European and other countries advanced in their demographic transition (China, Japan and Korea), this phenomenon implies an even more important reduction in investment.

Figure 3.6. The share of household consumption in GDP is projected to increase with ageing

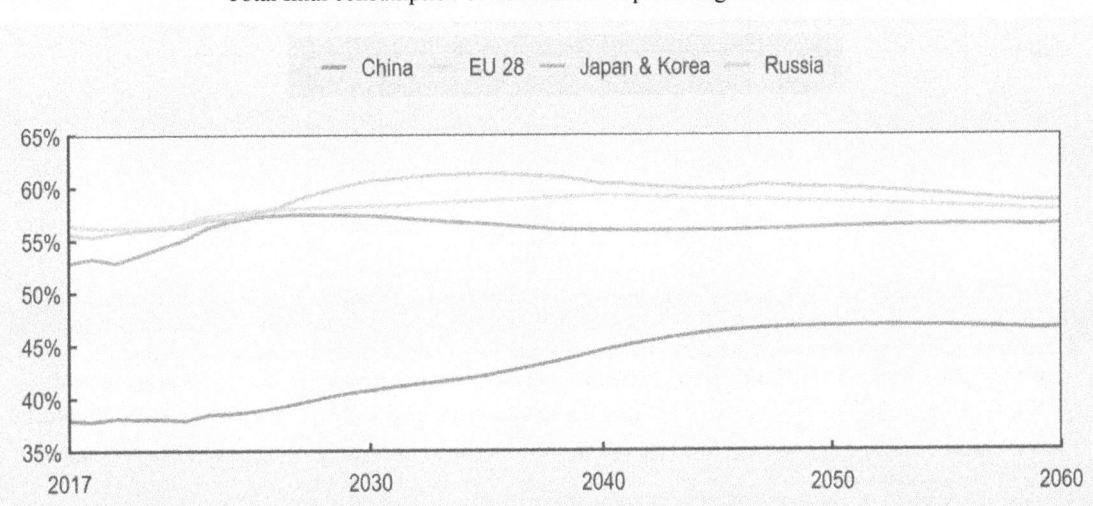

Total final consumption of households as percentage of total GDP

Source: OECD ENV-Linkages model; short-term forecasts by OECD Economics Department (as of Summer 2018) and IMF (as of Spring 2018).

StatLink https://doi.org/10.1787/888933884593

The direct consequence of this crowding out of investment by consumption is the growing share of demand for consumption goods as well as leisure and service activities and a reduction in demand for equipment and other goods that are used to build up the physical capital stock.

These trends imply a decoupling of materials use from GDP, as the sectors that are boosted by these trends are less materials-intensive than other sectors.

Figure 3.7. Labour efficiency and capital supply drive per-capita GDP growth

Annual growth rates in percentages

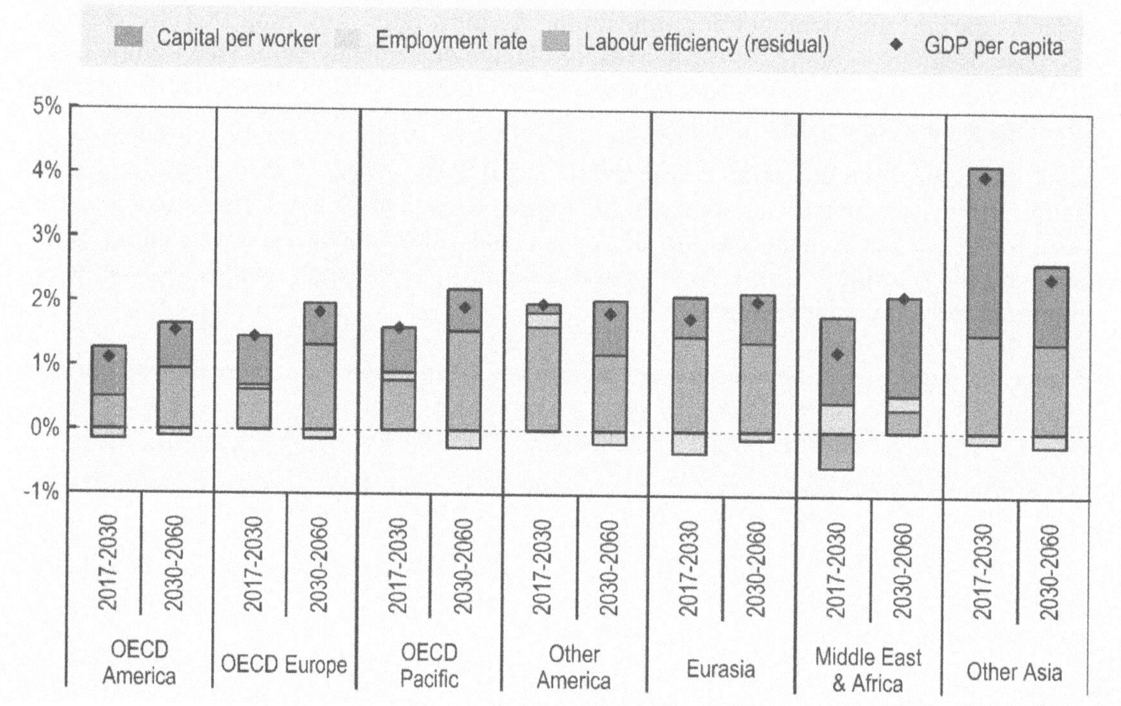

Note: The changes in the GDP per capita in market exchange rates (y) are decomposed in three components: (i) the change in employment rate (ER), (ii) the change in capital per worker (where capital is defined in a broad way including land and natural resources) (k), and, as a residual factor, (iii) the change in labour efficiency (A). Changes in GDP (in market exchange rates) can be decomposed as in the following formula: $\frac{\Delta y}{y} = \alpha \frac{\Delta A}{A} + (1-\alpha) \frac{\Delta k}{k} + \frac{\Delta ER}{ER}$, where α is the share of labour income in GDP. The GDP per capita growth rate in market exchange rates differs from the one in PPP exchange rates as the weights of different countries in regional aggregates differ.

Source: ENV-Growth model (OECD Environment Directorate) and OECD Economics Department (Guillemette and Turner, 2018[3]).

StatLink https://doi.org/10.1787/888933884612

Economic growth is thus characterised by changes in production technologies, which drive changes in the input structure (e.g. substitution of production inputs, labour or capital).[12,13] Such shifts in the input structure of production are not new – during the industrial revolution, for example, machines used to automate production reduced the need for labour. More recently, the increasing efficiency of cars has led to a lower use of fuel to travel the same distance, as well as a substitution between different types of fuels (e.g. ethanol instead of gasoline).

The production of manufacturing goods is an interesting example of these production changes. Table 3.1 illustrates changes over time in the cost structure of aggregate manufacturing good production, for OECD and non-OECD countries. Inputs of services increase, reflecting the servitisation phenomenon described in Section 3.2, while other inputs of goods and services –including extracted materials – decrease. Labour costs also increase, due to wage increases relative to the marginal cost of production (not shown here).

In both OECD and non-OECD countries, unit production costs are projected to decline, reflecting higher productivity resulting from technical progress. However, this effect is stronger in non-OECD countries, where a higher rate of convergence also leads to more marked changes in productivity over time. In all regions, production costs shift away from industrial inputs towards more services.

Table 3.1. Input composition for the production of manufacturing goods

Share of components in production costs of manufacturing goods

		OECD			Non-OECD		
		2017	2030	2060	2017	2030	2060
Price evolution (index 2017 = 1)		1.00	1.00	0.99	1.00	0.91	0.84
Input Composition of production	Capital and resources	11%	11%	12%	9%	9%	10%
	Labour	18%	19%	17%	14%	14%	14%
	Agricultural inputs	3%	4%	3%	8%	7%	7%
	Industrial inputs	47%	46%	40%	55%	54%	49%
	Services inputs	19%	21%	27%	14%	15%	21%

Source: OECD ENV-Linkages model.

StatLink https://doi.org/10.1787/888933885771

As new technologies emerge, are adopted and become cheaper, they will be more widely used for the production of goods. An example is electricity generation as electricity can be produced with different technologies. Over time renewable technologies are projected to become cheaper and easier to access so that they will be more widely used to produce electricity. In the central baseline scenario, which projects a gradual shift towards renewables, the percentage of electricity produced with renewable technologies is projected to increase at the global level from 24% in 2017 to 31% in 2040, while fossil fuel electricity is projected to decline from 65% in 2017 to 61% in 2040 (IEA, 2017[7]).

3.4. Several areas of uncertainty affect the socioeconomic projections

Several uncertainties need to be kept in mind when evaluating the materials projections in later chapters of this report: the central baseline scenario is carefully calibrated to reflect plausible developments, but represents only one possible pathway and is thus not a prediction of the future. This section presents alternative scenarios to identify which trends in the projections are robust, and which hinge critically on specific assumptions about the socioeconomic drivers.

Two key uncertainties are explored:

1. (i) Changes in population: alternative assumptions about population growth are modelled on the low and high population scenarios from the UN population prospects (2017[1]);

2. (ii) Changes in the speed of income convergence across countries: alternative assumptions about convergence are implemented by adjusting the speed with which labour efficiency of countries grows towards their country-specific potential.

Modelling these alternative assumptions leads to GDP variations from the central baseline scenario of -3% to +3% in 2030 and of -19% to +18% in 2060 respectively, as shown in

Figure 3.8. A key reason for the larger uncertainty around income convergence than population is the difference in timing: convergence differences have an immediate effect and lead to diverging GDP growth paths early on. In contrast, the variation in GDP from variation in population growth plays out over a much longer time horizon as there is a delay between population growth and growth in labour supply. The uncertainties identified here, which represent an uncertainty range of 306 to 444 trillion USD globally in 2060, translate into uncertainties about the projected use of materials, although not necessarily proportionately, as the changes in socioeconomic drivers may also affect the structure of the economy and material intensities.

Figure 3.8. Population and income convergence assumptions lead to long-term uncertainties in GDP

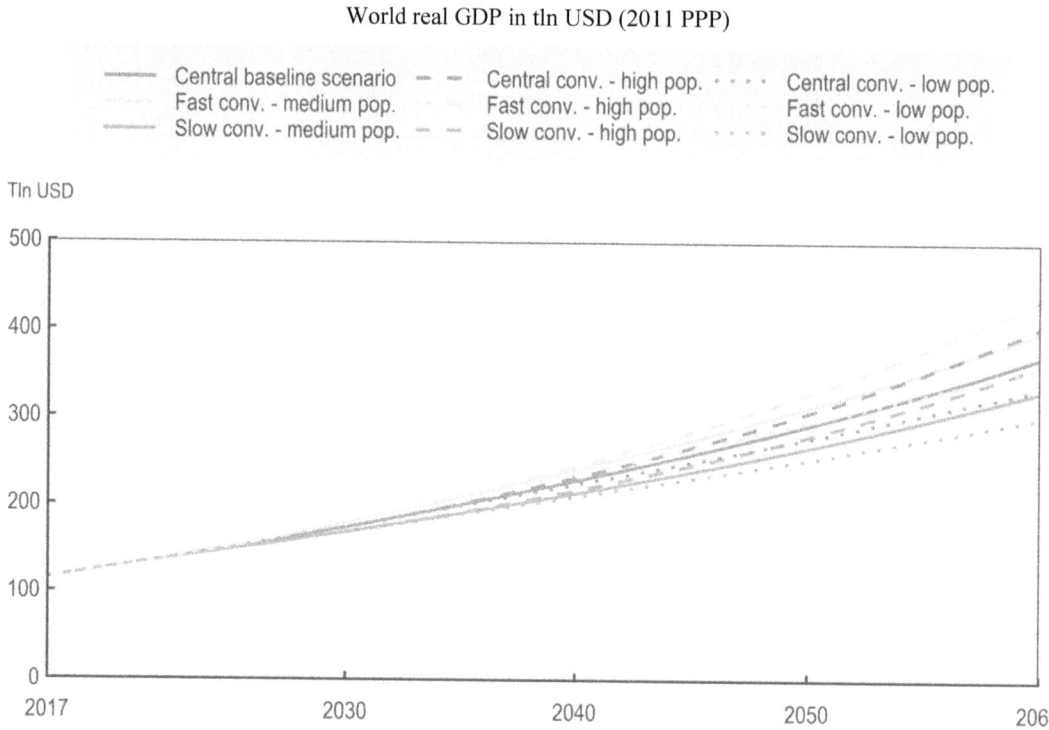

Note: The solid lines relate to central assumptions about economic convergence, the dotted lines show faster convergence assumption while the dashed lines show slower convergence. Blue lines relate to central population prospects (i.e. UN (2017[1]) medium scenario), the grey lines show the UN high population scenario and the green ones the UN low population scenario. The central blue line thus corresponds to the central baseline scenario.
Source: UN (2017[1]) population scenarios; ENV-Growth model (OECD Environment Directorate) and OECD Economics Department (Guillemette and Turner, 2018[3]).

StatLink https://doi.org/10.1787/888933884631

Panel A of Figure 3.9 shows the corresponding regional ranges. The absolute range in the Other Asia region, which comprises the non-OECD Asian countries, is projected to be significantly larger than in other regions: a variation of 40 trillion USD in 2060 for the macroeconomic convergence, and 74 trillion USD when adding the population assumptions. In comparison, the variations for the rest of the world combined are projected to reach 28 and 64 trillion USD respectively.

In relative terms, i.e. variations from the central baseline scenario, the uncertainty levels are projected to be much more similar across regions (Figure 3.9, Panel B). The non-OECD Asia region ("Other Asia") still has the largest uncertainty, given its high share of emerging economies which are heavily affected by assumptions about the speed of convergence. Most of the uncertainty is projected to derive from the convergence assumption rather than the population assumption, especially in non-OECD countries.[14]

Figure 3.9. Uncertainties surrounding population and income convergence assumptions are greatest for emerging economies

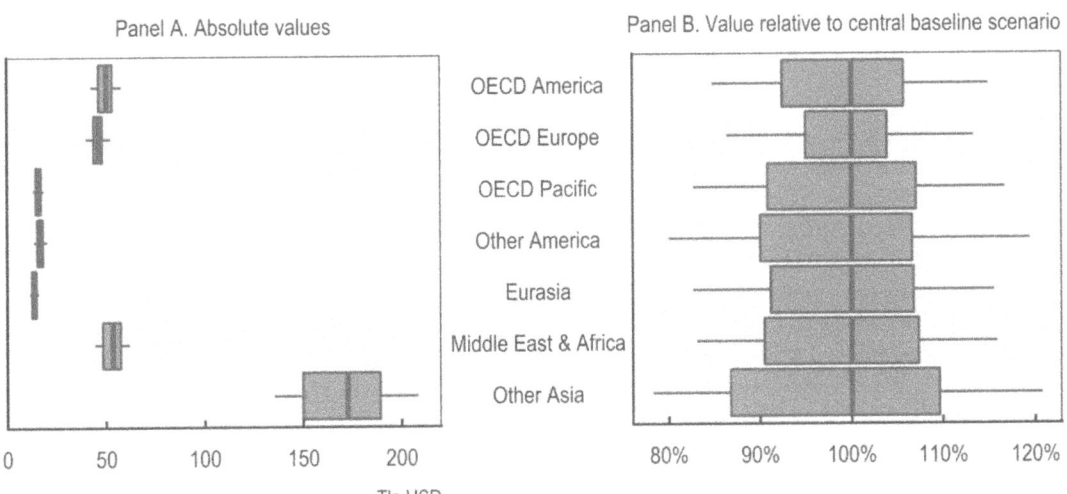

Note: The vertical black line shows the central baseline scenario, the size of the blue box corresponds to the level of uncertainty from the convergence assumptions, and the length of the whiskers the level of uncertainty from the population assumptions, superimposed on the convergence assumptions.
Source: UN (2017[1]) population scenarios; ENV-Growth model (OECD Environment Directorate) and OECD Economics Department (Guillemette and Turner, 2018[3]).

StatLink https://doi.org/10.1787/888933884650

Most of the variation in GDP comes from changes in employment, driven by the population assumptions, and from changes in labour productivity, driven by the convergence assumptions, as shown in Table 3.2. Capital changes following marginal changes in savings rates driven by changes in population structure are projected to be a lower source of uncertainty. The table also shows that the uncertainties are especially large for emerging and developing economies. These countries start from a lower level but are in the process of ramping up their economic growth; a change in the speed of convergence will affect them more strongly. In contrast, according to the UN population scenarios used here, the impacts of variations in the population projections are more equally shared across countries.

Table 3.2. Uncertainty surrounding the main macroeconomic variables in 2060

Percentage deviation from the central baseline scenario

		GDP	GDP per capita	Capital to GDP	Labour productivity	Employment	Population
OECD	Low Population & Slow convergence	-15%	-1%	7%	-7%	-11%	-14%
	Low Population & central convergence	-9%	6%	4%	3%	-11%	-14%
	Low Population & Fast convergence	-4%	11%	2%	11%	-11%	-14%
	Medium Population & Slow convergence	-7%	-7%	3%	-10%	0%	0%
	Medium Population & Fast convergence	5%	5%	-2%	8%	0%	0%
	High Population & Slow convergence	1%	-12%	0%	-12%	11%	16%
	High Population & central convergence	9%	-6%	-3%	-2%	11%	16%
	High Population & Fast convergence	15%	-1%	-5%	5%	11%	16%
Non-OECD	Low Population & Slow convergence	-20%	-6%	7%	-11%	-11%	-15%
	Low Population & central convergence	-10%	6%	5%	1%	-11%	-15%
	Low Population & Fast convergence	-2%	15%	4%	11%	-11%	-15%
	Medium Population & Slow convergence	-12%	-12%	3%	-13%	0%	0%
	Medium Population & Fast convergence	9%	9%	-1%	10%	0%	0%
	High Population & Slow convergence	-4%	-17%	-1%	-14%	11%	16%
	High Population & central convergence	10%	-6%	-4%	-1%	11%	16%
	High Population & Fast convergence	19%	2%	-4%	9%	11%	16%
World	Low Population & Slow convergence	-19%	-5%	6%	-9%	-11%	-15%
	Low Population & central convergence	-10%	6%	4%	2%	-11%	-15%
	Low Population & Fast convergence	-3%	14%	3%	11%	-11%	-15%
	Medium Population & Slow convergence	-10%	-10%	2%	-11%	0%	0%
	Medium Population & Fast convergence	8%	8%	-1%	9%	0%	0%
	High Population & Slow convergence	-2%	-16%	-1%	-13%	11%	16%
	High Population & central convergence	9%	-6%	-3%	-2%	11%	16%
	High Population & Fast convergence	18%	1%	-4%	7%	11%	16%

Note: GDP numbers are in real terms, expressed in 2011 USD at PPP exchanges rates. Labour productivity is defined here as the ratio of employment to GDP.
Source: UN (2017[1]) population scenarios; ENV-Growth model (OECD Environment Directorate) and OECD Economics Department (Guillemette and Turner, 2018[3]).

StatLink https://doi.org/10.1787/888933885790

These uncertainty ranges highlight how relatively small variations in the socioeconomic drivers can have substantial consequences for future levels of GDP. The dynamic linkages in the economy, whereby a short-term boost of the growth rate can lift a country onto a faster growth path, thereby boosting future GDP levels, have a significant effect on the projections. The "true" assumptions regarding population growth and convergence speed cannot be determined ex ante. Therefore, these uncertainties need to be kept in mind when evaluating the materials projections in later chapters of this report: the central baseline scenario is carefully calibrated to reflect plausible developments, but represents only one possible pathway and is thus not a prediction of the future. The alternative scenarios presented in this section are thus useful to identify which trends in the projections are robust, and which hinge critically on specific assumptions on the socioeconomic drivers.

Notes

[1] Income demography are connected in reality, but the dynamics of both components is discussed separately for simplicity.

[2] See detailed assumptions in Table 3.A.1 in Annex 3.A.

[3] In the simulations, South Africa is separated from the rest of Sub-Saharan Africa, which is labelled "Other Africa". In the text, the term Other Africa is replaced with Sub-Saharan Africa to improve readability.

[4] Other EU includes the non-OECD EU countries (see Table 2.1 in Chapter 2).

[5] Details by region are provided in Table 2.A.1 in Annex 3.A .

[6] In the ENV-Linkages model, services are split into several sectors: business services, three transport service sectors (land, air, water), and other services (which include all government services, education, health, waste management).

[7] To illustrate these trends, (Miroudot and Cadestin, 2017[8]) showed that in 2015 in OECD countries between 25% and 60% of employment in manufacturing firms was in service support functions such as R&D, engineering, transport, logistics, distribution, marketing, sales, after-sale services, IT, management and back-office support.

[8] Energy intensive industries include the sectors producing chemicals, iron & steel, pulp, paper publishing and mon-metallic minerals.

[9] Annex 3.A provides more precise elements about the decomposition of GDP per capita growth into these three drivers and about the underlying assumptions about the changes in those drivers.

[10] The methodology for projecting labour efficiency has been developed by the OECD Economics Department. Guillemette et al. (2017[9]) describe this methodology as well as the projection for the underlying determinants of long run efficiency. For the remaining 180 countries, the OECD Environment Directorate adopts a similar methodology but with fewer determinants in the long run efficiency: indicators for institutional quality as well as rule of law are not included in ENV-Growth model due to lack of data.

[11] For standard growth models (e.g. the Solow-Swan growth model), the capital-to-output ratio and the employment rate stabilise in the long run following the convergence towards a balanced growth path where capital supply growth matches labour efficiency growth.

[12] These effects are driven by changes in relative costs and in factor productivity that affect the mix of inputs and technologies used to produce the final goods. The input substitution effect occurs as the price of one input changes relative to other inputs. In particular, if different inputs can serve as substitutes in the production of a specific commodity, then the mix of inputs that are used for the production of this commodity will depend on their relative prices. Further, if production inputs become more efficient through increases in total factor productivity, then more output can be generated with the same amount of inputs.

[13] In the modelling framework, smooth production functions are used to represent the production choices of many individual firms. Individual production technologies are only specified for selected sectors, notably those related to energy production and materials processing.

[14] The downward variation from the reduced speed of convergence has a larger impact on future GDP levels than the increased convergence speed. The main reason for this is that an equal

downward shift in the speed of convergence has a larger impact on absolute levels than the same rate of upwards shift.

References

de la Maisonneuve, C. and J. Oliveira Martins (2014), "The future of health and long-term care spending", *OECD Journal: Economic Studies*, Vol. 2014/1, http://dx.doi.org/10.1787/eco_studies-2014-5jz0v44s66nw. [4]

Eurostat (2018), "Population projections", *Eurostat (online data code: tps00002)*, http://ec.europa.eu/eurostat/web/products-datasets/-/tps00002 (accessed on July 2018). [2]

Guillemette, Y. et al. (2017), "A revised approach to productivity convergence in long-term scenarios", *OECD Economics Department Working Papers*, No. 1385, OECD Publishing, Paris, http://dx.doi.org/10.1787/0b8947e3-en. [9]

Guillemette, Y. and D. Turner (2018), "The Long View: Scenarios for the World Economy to 2060", *OECD Economic Policy Papers*, No. 22, OECD Publishing, Paris, http://dx.doi.org/10.1787/b4f4e03e-en. [3]

IEA (2017), *World Energy Outlook 2017*, OECD Publishing, Paris/IEA, Paris, http://dx.doi.org/10.1787/weo-2017-en. [7]

IMF (2004), "How will demographic change affect the global economy?", in *World Economic Outlook: The Global Demographic Transition*, World Economic and Financial Surveys, Washington DC, http://www.imf.org/external/pubs/ft/weo/2004/02/. [6]

Miroudot, S. and C. Cadestin (2017), "Services In Global Value Chains: From Inputs to Value-Creating Activities", *OECD Trade Policy Papers*, No. 197, OECD Publishing, Paris, http://dx.doi.org/10.1787/465f0d8b-en. [8]

Pilat, D. and A. Nolan (2016), "Benefiting from the next production revolution", in Love, P. (ed.), *Debate the Issues: New Approaches to Economic Challenges*, OECD Publishing, Paris, http://dx.doi.org/10.1787/9789264264687-22-en. [5]

UN (2017), "World Population Prospects: key findings and advance tables", https://esa.un.org/unpd/wpp/publications/Files/WPP2017_KeyFindings.pdf (accessed on 18 May 2018). [1]

Annex 3.A. Detailed results and supplementary materials

3.A.1 Detailed total population projections and the ageing process

Global population growth is projected to slow down. This global trend results from a decrease in population growth in all regions, including both OECD and non-OECD countries (see Table 3.A.1).

Table 3.A.1. Population by region: historical and projected trends

	Average annual growth		Percentage of world total	
	1980-2020	2020-2060	2000	2060
World	1.4%	0.7%	100%	100%
Japan	0.2%	-0.5%	2.1%	1.0%
Korea	0.8%	-0.2%	0.8%	0.5%
OECD Oceania countries	1.3%	0.8%	0.4%	0.4%
Canada	1.1%	0.5%	0.5%	0.4%
Chile	1.3%	0.3%	0.2%	0.2%
Mexico	1.7%	0.5%	1.7%	1.6%
United States of America	0.9%	0.5%	4.6%	4.0%
European Union OECD 17 Smaller countries	0.3%	-0.2%	3.1%	1.8%
European Union OECD 4 Larger countries	0.3%	0.0%	4.3%	2.7%
European Union Non OECD countries	-0.4%	-0.7%	0.7%	0.3%
Other OECD Eurasian countries	1.5%	0.5%	1.3%	1.3%
Non-EU Eastern Europe countries	-0.2%	-0.7%	1.3%	0.6%
Russia	0.1%	-0.3%	2.4%	1.2%
Caspian countries	1.2%	0.5%	1.2%	1.1%
Middle East countries	2.6%	1.1%	2.7%	3.8%
North African countries	1.9%	1.0%	2.4%	2.9%
Other African countries	2.8%	2.2%	10.2%	25.0%
South Africa	1.7%	0.4%	0.7%	0.7%
Other ASEAN countries	1.6%	0.5%	5.2%	4.7%
China	0.9%	-0.2%	20.9%	12.6%
Indonesia	1.5%	0.5%	3.5%	3.2%
India	1.7%	0.6%	17.3%	17.2%
Other Developing Asia countries	2.0%	0.8%	6.0%	6.9%
Brazil	1.4%	0.2%	2.9%	2.3%
Other Latin America countries	1.5%	0.5%	3.8%	3.6%

Source:
Own calculation from The World Population Prospects: 2017 Revision (UN, 2017[1]) and Eurostat (Eurostat, 2018[2]).

The slowdown in population growth is mostly due to ageing. The ageing process is ongoing at both extremities of the age pyramid (i.e., children and seniors). Indeed the left panel of Figure 3.A.1 shows that the senior population is projected to increase faster in the 2020-2030 period than in the past, due to the ageing of the numerous baby-boomer cohorts. After 2030, the increase is still substantial but at a slower pace.

The share of children in total population is projected to decrease everywhere (right panel of Figure 3.A.1). As a result, there will be fewer and fewer new entrants to the active population in the next decades. A more detailed analysis shows that the total number of children is going to decrease in 2060 relative to the actual level in most countries. Only Middle East & Africa, North America and Oceania project an increase of this population over the next four decades.

Figure 3.A.1. Shares of children and elderly in total population

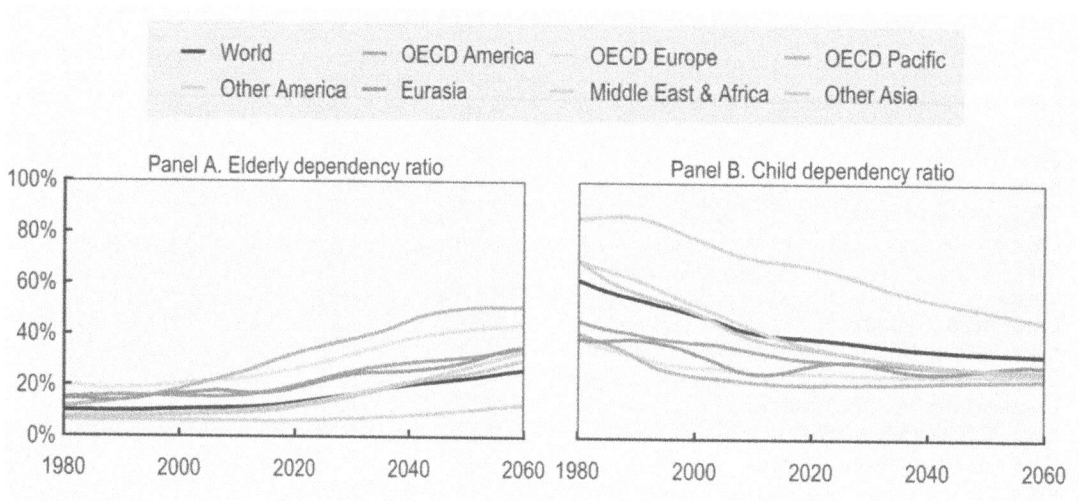

Note: *Elderly dependency ratio* is population over 65 years old to population between 15 and 65 years old; *Child dependency ratio* is population younger than 15 years old to population between 15 and 65 years old.
Source: Own calculations from The World Population Prospects: 2017 Revision (UN, 2017[1]).

The increasing share of elderly people and the decline in the number of children lead to the projected decrease (shown in Figure 3.A.2) in growth rate of the working-age population for the period 2020-2040 (relative to the period 1980-2020.

In the last 40 years globally, the working age population accounted for almost 1.7% increase per year; in the next four decades, the same indicator is projected to fall to a modest 0.7%/year, this slow down for working age population is therefore more pronounced than that of the total population. Most European countries, Japan and Korea, currently observe a decline in their working age population, and this decline is projected to accelerate in the coming years, and then stabilize around 2040 at about -0.25% per year. Projections for China and Russia exhibit a similar profile, while for OECD America and Oceania, the growth is 0.4% per year. In contrast, many African countries observe a 2% growth per year in active population. Projected profiles for other countries are close to the world average.

Figure 3.A.2. Growth of the working age population

Annual average growth rate of population from 15 to 75 years old

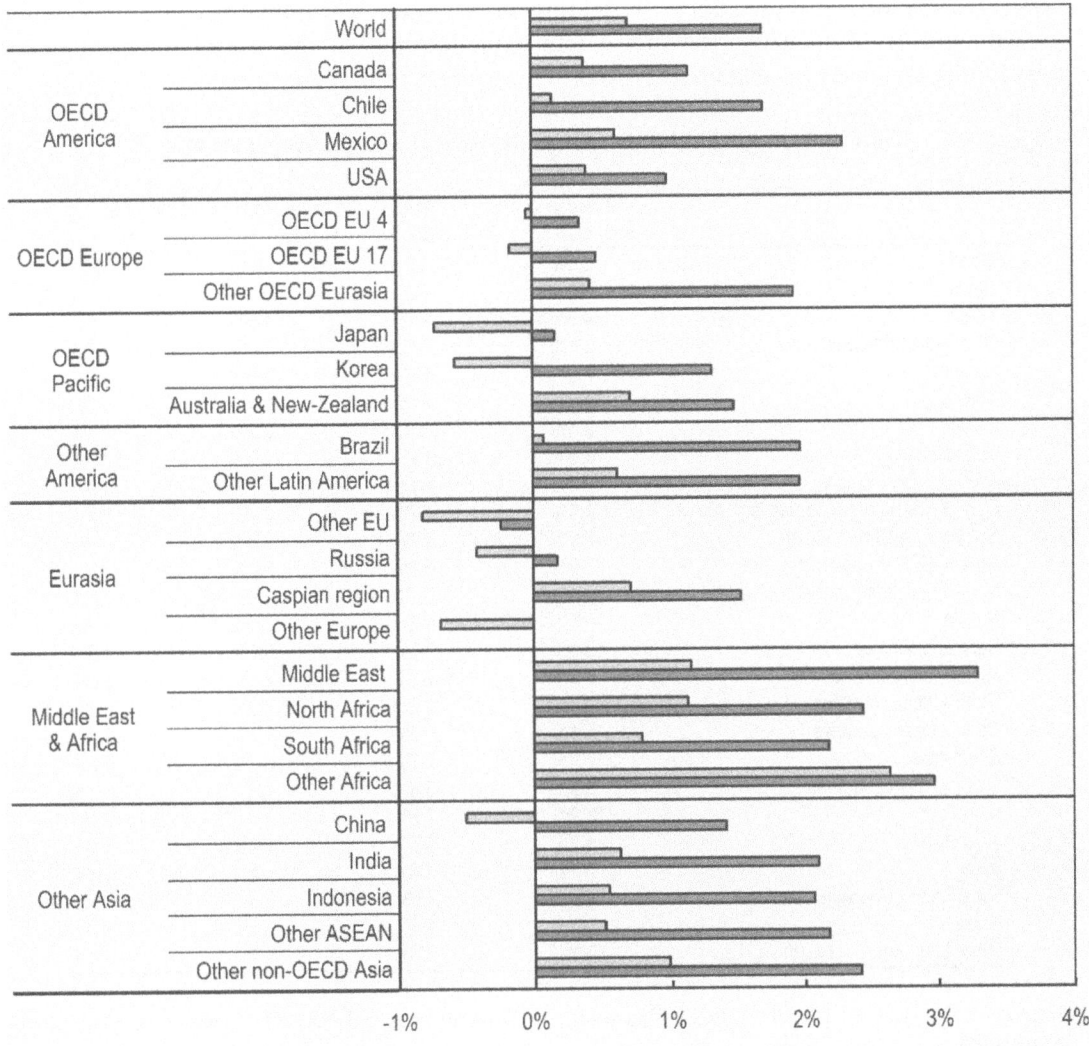

Source: Own calculation from The World Population Prospects: 2017 Revision (UN, 2017[1]).

3.A.2 Detailed GDP growth and assumptions about drivers of the projected GDP per capita

GDP growth is projected to stabilise at global level (see Table 2.A.1). However, this is the result of uneven changes in GDP growth in different regions. In most OECD countries GDP growth will remain stable or decline in the long-term. In emerging economies growth will be high in the short-term and then decrease in the longer term. In many developing countries instead GDP growth will increase in the longer run.

Table 3.A.2. Real GDP by region: historical and projected trends

	Average annual growth		Percentage of world total	
	1990-2020	2020-2060	2000	2060
World	3.5%	2.8%	100%	100%
Japan	1.0%	1.2%	6.7%	2.1%
Korea	4.9%	1.7%	1.6%	1.1%
OECD Oceania countries	3.1%	2.7%	1.2%	1.0%
Canada	2.5%	1.9%	1.9%	1.0%
Chile	4.6%	2.0%	0.4%	0.3%
Mexico	2.3%	2.5%	2.6%	1.7%
United States of America	2.5%	1.9%	21.2%	10.6%
EU - OECD 17 Smaller countries	2.3%	1.7%	8.1%	3.9%
EU - OECD 4 Larger countries	1.5%	1.7%	14.4%	5.8%
EU - Non OECD countries	2.9%	2.1%	0.8%	0.6%
Other OECD Eurasian countries	3.9%	2.9%	2.6%	2.8%
Non-EU Eastern European countries	0.8%	3.1%	0.8%	0.7%
Russia	-0.4%	0.8%	3.1%	1.3%
Caspian countries	3.0%	3.7%	0.5%	1.3%
Middle East countries	4.0%	2.9%	4.1%	5.0%
North African countries	3.3%	4.2%	1.9%	3.1%
Other African countries	4.4%	5.3%	1.7%	6.5%
South Africa	2.7%	2.7%	0.7%	0.6%
Other ASEAN countries	5.2%	3.6%	2.9%	5.2%
China	8.9%	2.3%	7.8%	16.7%
Indonesia	5.1%	3.6%	1.9%	3.5%
India	7.0%	4.4%	4.4%	16.4%
Other Developing Asian countries	4.8%	4.0%	2.3%	4.4%
Brazil	5.6%	1.8%	3.1%	1.7%
Other Latin America countries	3.6%	2.8%	3.5%	3.0%

Source: ENV-Growth model (OECD Environment Directorate) and OECD Economics Department (Guillemette and Turner, 2018[3]).

This evolution of GDP rests on the compounded evolution of populations and living standards. Figure 3.A.3 shows the evolution of living standards between 2017 and 2060 as measured by GDP per capita.

For making long-term projections, GDP per capita growth is a common indicator of economic trends, but it does not explain the sources and the drivers of economic growth. It is therefore worthwhile to isolate the underlying drivers of GDP per capita. A first step to explain GDP per capita consists in comparing projected trends tin the employment rates (measured as the share of employment in total population) and trends in labour productivity (measured as GDP per person employed).

Then a second step consists in isolating the internal drivers of changes in employment rates and then the drivers of changes in GDP per worker (which includes the effects of other drivers of growth, such as capital deepening).

Figure 3.A.3. GDP per capita evolution between 2017 and 2060

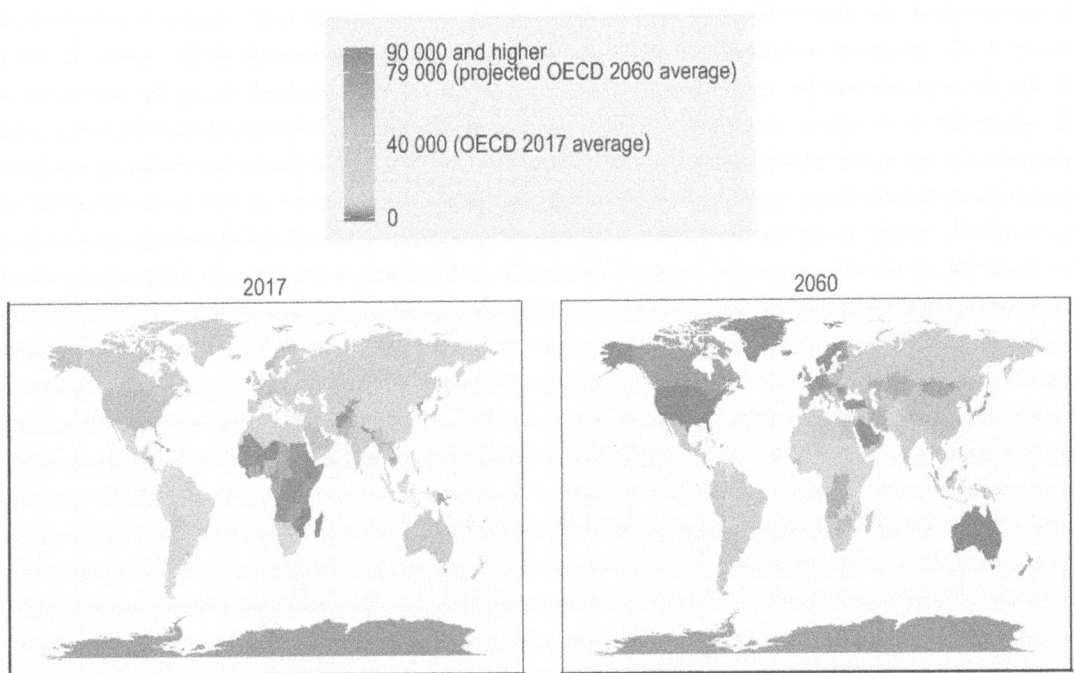

Source: ENV-Growth model (OECD Environment Directorate) and OECD Economics Department (Guillemette and Turner, 2018[3])

Projections of employment rates

Employment rates are projected to change in the future according to three main components: (i) the structure of the population (defined as the share of working age population in total population), (ii) the labour participation rate (defined as the share of active population to working age population) and (iii) the unemployment rate (the share of unemployed to active population).

Figure 3.A.4 indicates a negative contribution of employment rates to GDP per capita (diamond mark in the figure) at world level (last column), over the projection period (2017-2030). Looking more deeply at the results shows that the unemployment rates are projected to decline or remain constant in almost all regions but Africa, Middle East and in some part of Asia and Latin America. The bars in Figure 3.A.4 show the contributions of the three components to employment rate. For ageing countries like OECD countries and Russia, the decline in employment rates is largely but not entirely attributable to the changes in population structure (e.g. the reduction of the size of working age population to total population). For these same countries, the participation rates are increasing (mostly women joining the workforce), and therefore partly offset the effect of ageing.

In emerging and developing economies characterized by a dynamic growth of both population and income (like many African economies), both changes in population structures and employment contribute positively to the increase in employment rates. For Asian countries, no common pattern could be highlighted. China, for example, is characterized by decrease in both the contributions of the population structure and the participation rates. Indonesia is projected to face exactly the opposite situation of both components growing, while for India the decline in the participation rate component overcomes the positive impact of the population structure component.

Figure 3.A.4. Changes in the drivers of employment rates

Annual average growth rate of employment to total population, 2017-2060

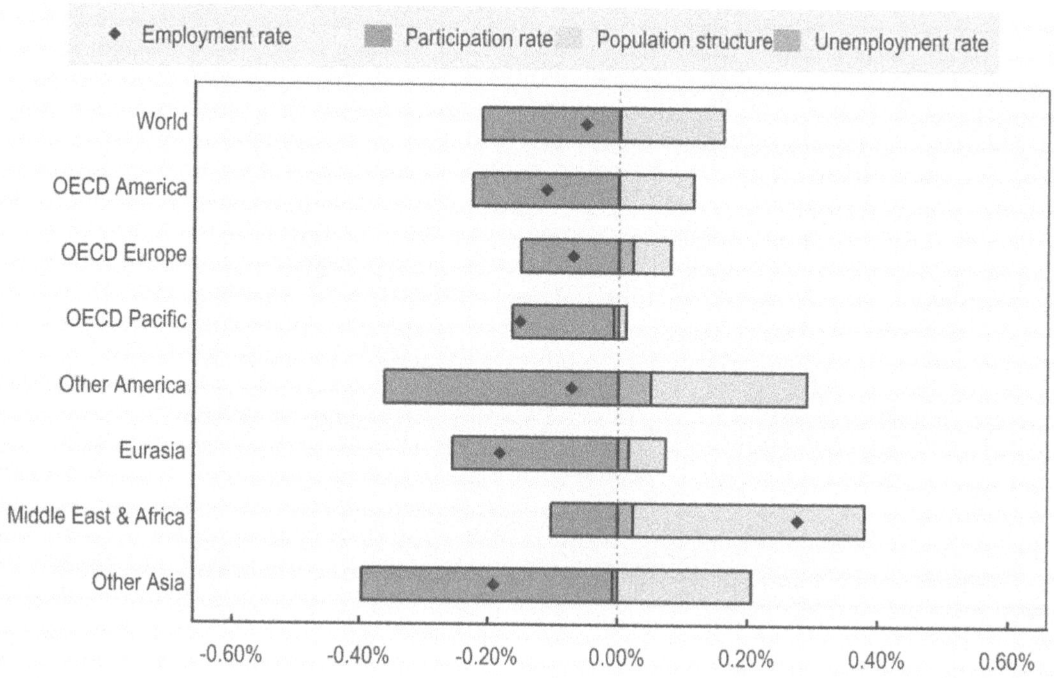

Note: The changes in the employment rate (ER) is mechanically decomposed in the sum of three components: (i) changes in the population structure, defined as the share in working-age population in total population (WR); (ii) changes in labour participation rate (PR); (iii) changes in unemployment rate (UR). Hence, Δ (ER) / ER = Δ (PR) / PR + Δ (WR) / WR − [UR/(1-UR)] . Δ (UR) / UR.
Source: ENV-Growth model (OECD Environment Directorate) and OECD Economics Department (Guillemette and Turner, 2018[3]).

Projected trends in GDP per worker

In the long run, labour productivity, i.e. GDP per worker, accrues from labour efficiency improvements, as well as capital deepening, including land and other natural resources use. Over the projection period, the productive capital stock follows projected investment in physical capital (such as building, machines and equipment). The latter is mostly driven by savings (where demographic changes play a central role), but not only, since the assumptions about current account unbalances include to partly dissociate investment from national savings (see Figure 3.A.5).

As shown in Figure 2.A.1 the projection framework assumes that in the long run the physical capital stock and the GDP will increase at the same pace. In the medium run capital generally increases faster than GDP. This is either because it is necessary to invest in new capital to match the growth in labour input (in regions where employment is growing fast) or to adjust to the gains of technical progress.

Figure 3.A.5. Evolution of net-investment financial components

Billions of USD (2011 PPP)

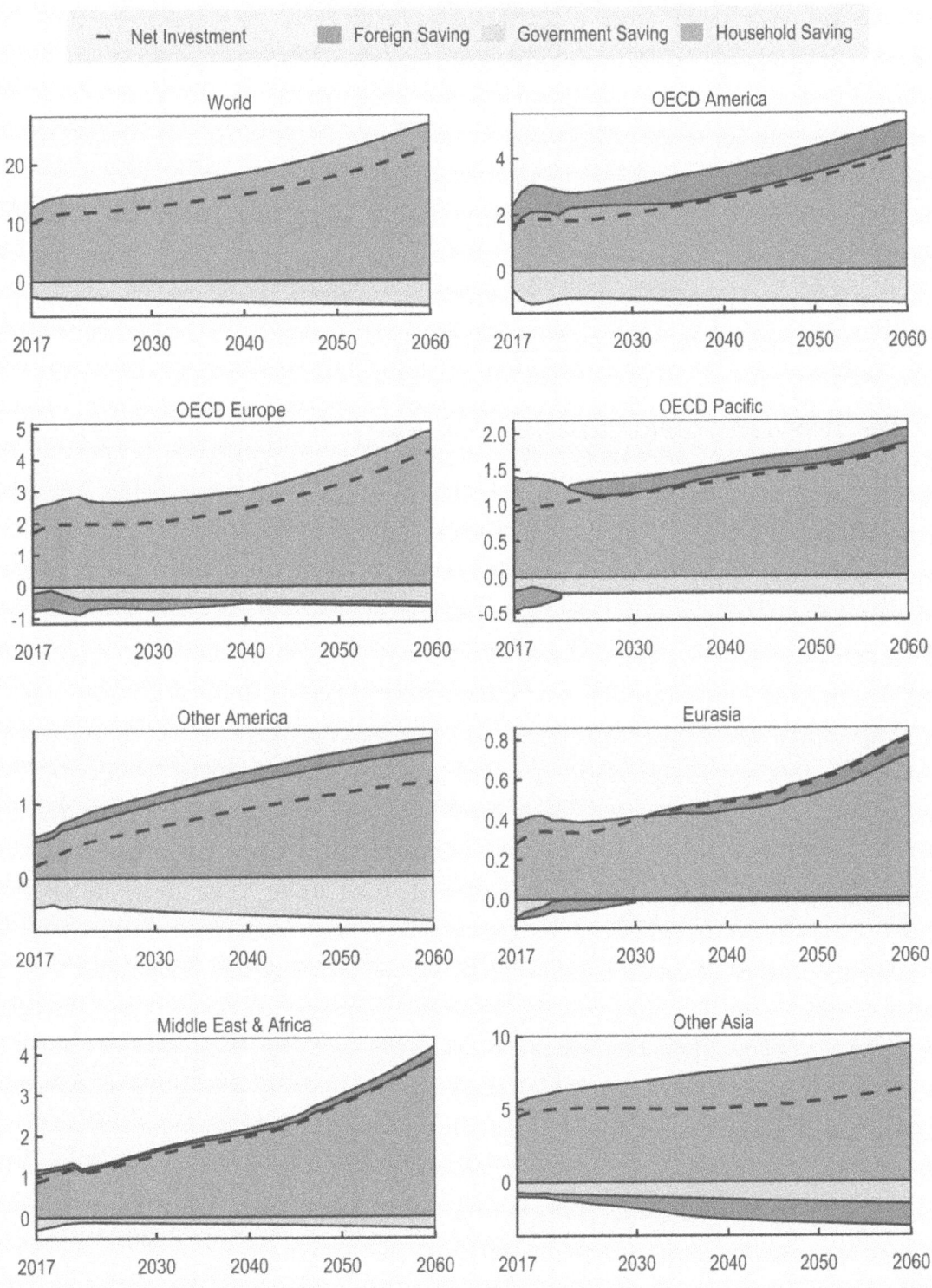

Note: The gap between investment and saving is the current account.
Source: ENV-Growth model (OECD Environment Directorate).

Figure 3.A.6. Evolution of capital to GDP ratios

Physical capital stock to real GDP, 2017-2060

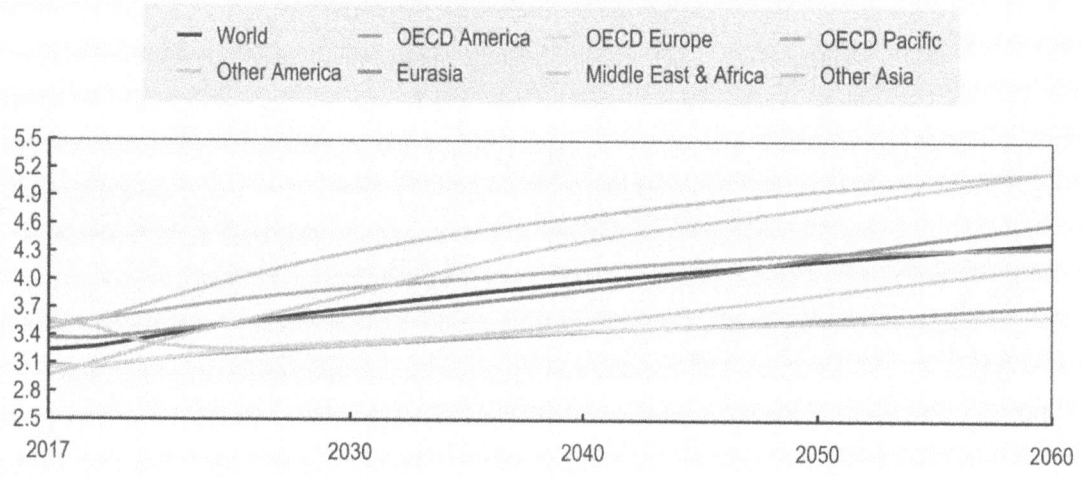

Source: ENV-Growth model (OECD Environment Directorate).

Chapter 4.

Projections of the economic drivers of materials use

This chapter explains how the projected changes in economic activities drive the use of primary materials and secondary materials. This chapter first explains how economic growth drives investment in physical capital, and thus the demand for construction materials. Then, it shows how changes in demand patterns, and in particular the growing share of services in the economy, can slow down materials use. Next, this chapter shows how changes in production processes and technologies affect materials intensity. This chapter also addresses the issue of the saturation of demand for materials, focusing on a potential saturation of the demand for construction materials in China.

KEY MESSAGES

This chapter describes three main economic mechanisms that are projected to drive materials use to 2060: (i) the gradual economic catching up of emerging and developing economies to the current living standards of mature economies; (ii) the servitisation of the economy; (iii) changes in production modes. The resulting saturation effects in the construction sector, especially in China, are also explored.

Projections and trends

As the economies of fast-growing countries mature and they build up infrastructure, factories and housing, their use of materials (mostly non-metallic minerals and metals) increases strongly. After the investment boom, materials use tends to stabilise and is directed mostly at investment that replaces existing infrastructure, which tends to involve less intense use of materials.

- The projected economic growth in emerging economies, especially India, and later in Sub-Saharan Africa, drives an increase in construction materials use. Despite a slowdown in its economic growth, over the projection period materials use remains the highest in China.

- Globally, the construction sector more than doubles between 2017 and 2060, as does its use of materials, leading to almost 84 gigatonnes (Gt) construction materials use per year in 2060.

Figure 4.1. Global growth is faster in less materials-intense sectors

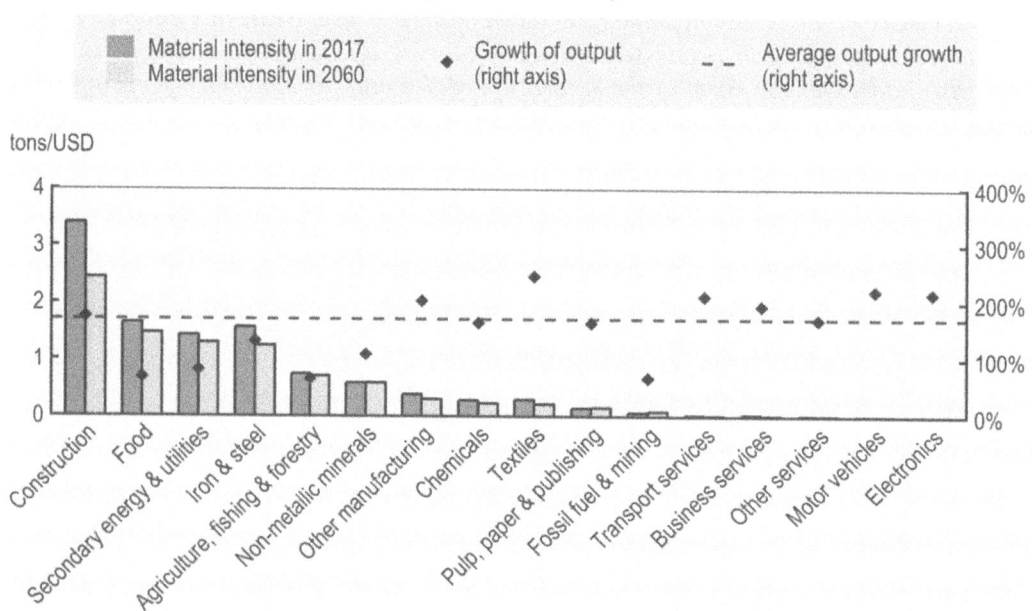

Note: Intensities are in tonnes per k USD. The bars for the sectors with very low values are too small to show on this graph (e.g. Electronics).
Source: OECD ENV-Linkages model.

StatLink https://doi.org/10.1787/888933884669

- The increasing share of services in manufacturing and demand by households and government, combined with other trends such as digitalisation and an increase in R&D, enhance the share of the services sectors in the economy. This means that global materials intensity decreases over the projected period due to the relatively low materials intensity of services compared to agriculture and industry (see Figure 4.1). Despite this projected "relative decoupling", materials-intensive sectors continue to grow until 2060, leading to a substantial increase in overall materials use. For instance, global demand for food and agricultural goods is projected to increase by about 65% by 2060 over 2017 levels.

- For metals, the growth in global demand and the shift of production towards regions with more materials-intensive production sectors are in the short to medium run projected to outweigh the improvements in materials productivity at the sectoral level. But in the longer run the productivity improvements are projected to lead to a (relative) decoupling of primary materials use in metals production.

Areas of uncertainty

Developments in China have a huge influence on global materials use, with China's construction sector being the biggest single source of materials use projected in the model. How materials use grows in the future partially depends on whether the demand for construction materials eventually stabilizes as housing needs are fulfilled.

The maturing of the Chinese economy is projected to lead to a slowdown of materials use in relation to the growth in GDP thanks to a saturation of demand, as well as changes in the materials used for building (less cement).

The chapter presents alternative baseline projections of materials use. While the central baseline scenario has construction materials use in China stabilising below 25 Gt per year, a scenario based on existing trends – excluding structural and technology trends – would lead to the use of 67 Gt of construction materials in China by 2060. This difference represents more than half of total global consumption of construction materials.

Policy implications

Quantifying the economic mechanisms that drive materials use is crucial for designing sound policies to promote resource efficiency and a transition to a circular economy. For instance, if the catching up of emerging and developing economies is a policy goal in itself, that objective may lead to increased materials use if no changes are made in production modes. A well-designed policy package will then be needed to further decouple growth from materials use.

4.1. Economic development and construction materials are closely linked

There is a strong link between economic development and materials use, as materials are an important input for all production processes. There is a particularly strong link between economic development, investment, construction activity and demand for construction materials, which is analysed in this section.

In line with historical developments, the economic baseline scenarios assume that income levels in less developed countries will catch up gradually – depending on country-specific characteristics – with more advanced economies; this conditional convergence process is laid out in Section 2.1. Often, these phases of catch up are characterised by a capital investment boom, as China has experienced in the last two decades. These investment booms result in a rapid increase in housing and infrastructure construction as well as in new machinery purchases.[1] In this context, as many emerging economies are still far from fully catching up to OECD living standards (Figure 3.3), a high level of investment in equipment, housing and infrastructure is projected to occur in the next decades.

The projected contribution of the different regions to global growth, defined as the share of global GDP growth originating in the region, is presented in Figure 4.2. The largest share of global growth to 2060 is projected for non-OECD countries, while the contribution to global GDP growth by OECD countries remains roughly constant around 45%. Countries with strong population growth (such as India and the Sub-Saharan African countries[2]) increase their contribution to global growth. India's contribution to global GDP growth is projected to increase from 7% in 2017 to 11% by 2060 and Sub-Saharan Africa's share rises from 2% to 11% over the same period.[3] In contrast, China's contribution to annual GDP growth is projected to decline from 30% in 2017 (i.e. almost one-third of current global GDP growth can be attributed to China) to 9% in 2060. As highlighted in Chapter 2, the overall global growth rate is projected to steadily fall in the coming decades, stabilising after 2030 to just below 2.5% per year.

Figure 4.2. Emerging Asian and Sub-Saharan African economies are projected to replace China as engines of global growth

Contribution of regions to global real GDP growth (percentage).

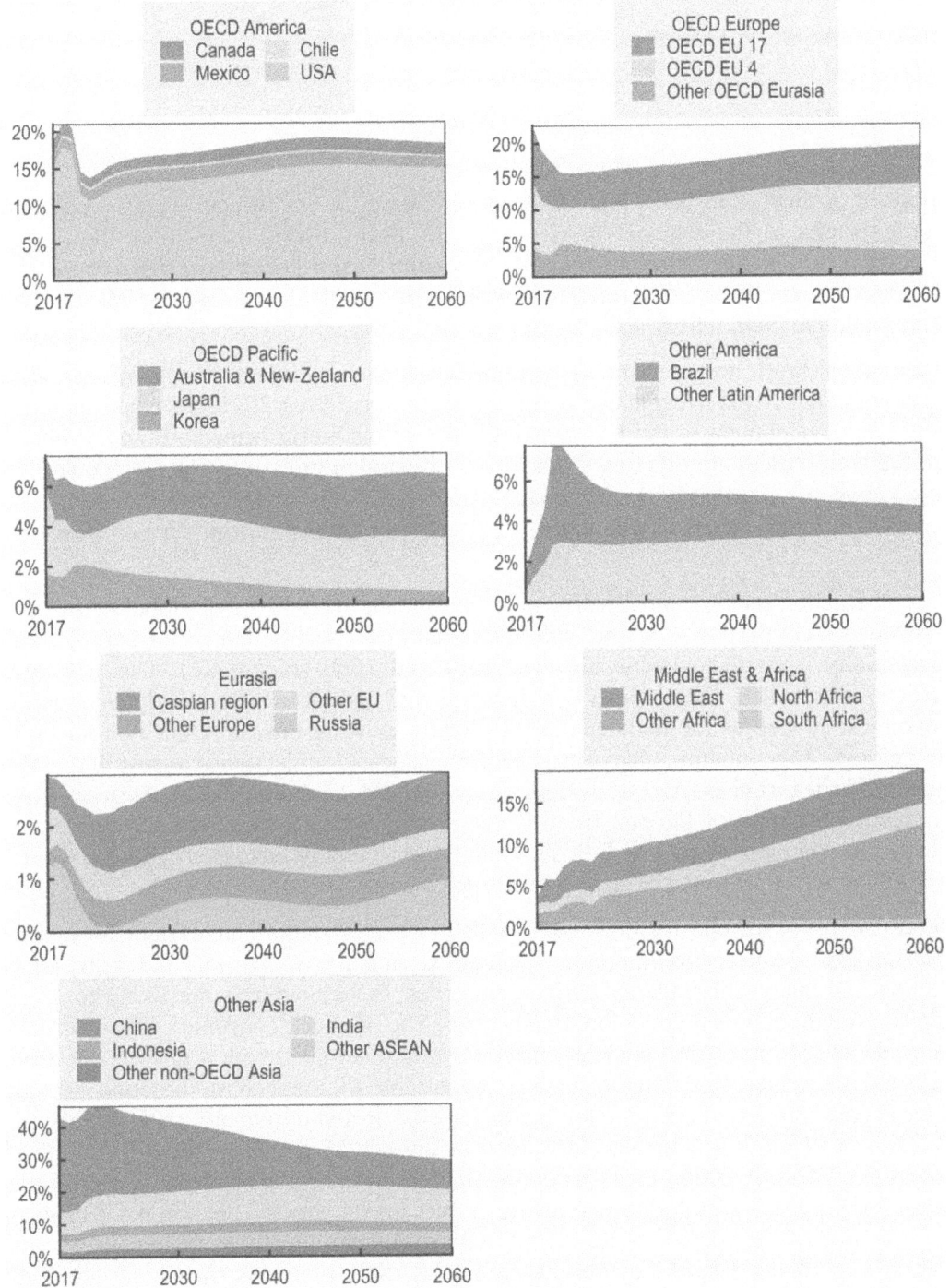

Notes:
1. GDP is measured as real GDP in constant 2011 USD using PPP exchange rates. The numbers in the graph add up to 100% across regions.
2. See Table 2.1 for regional definitions. In particular, OECD EU 4 includes France, Germany, Italy and the United Kingdom. OECD EU 17 includes the other 17 OECD EU member states. Other OECD Eurasia

includes the EFTA countries as well as Israel and Turkey. Other EU includes EU member states that are not OECD members. Other Europe includes non-OECD, non-EU European countries excluding Russia. Other Africa includes all of Sub-Saharan Africa excluding South Africa. Other non-OECD Asia includes non-OECD Asian countries excluding China, India, ASEAN and Caspian countries.

3. The projected spike in Brazil at the beginning of the next decade results from the combination of a projected decrease of GDP growth rate in all regions of the world with a catch up in Brazil from the low projected growth rates in the last few years of this decade.

Source: OECD ENV-Linkages model; short-term forecasts by OECD Economics Department (as of Summer 2018) and IMF (as of Spring 2018).

StatLink https://doi.org/10.1787/888933884688

Sustained growth translates into a steady increase in investment levels. Figure 4.3 highlights this at the regional level. Especially fast-growing economies have higher investment levels and thus rapid capital accumulation relative to GDP, while more advanced economies are characterized by a higher level of installed capital and more stable capital to GDP ratios.

At the global level, however, the increase of investment is projected to be lower than the increase in GDP. This is explained by the projection that after 2025, the consequences of economic development in the fast-growing parts of the world (especially India, but also Sub-Saharan Africa) are offset in terms of net investment growth by the slowdown in the global economy. Furthermore, the gradual reduction in the saving rates that result from increases in living standards and ageing will also put downward pressure on investment rates.

The construction sector is mostly driven by investment needs: 90% of global construction is for investment purposes. In line with investment, the construction sector is projected to substantially more than double between 2017 and 2060 (Figure 4.4). This substantial increase rests on the development of housing and infrastructure, mostly in emerging economies, as well as the maintenance of already existing infrastructure in both OECD and non-OECD economies. The central baseline scenario projections suggest that in Sub-Saharan Africa (Other Africa in the figure, which excludes South Africa), output (production volume) of the construction sector may increase twelvefold in the coming decades, and in most of the non-OECD Asian regions, it may increase fivefold, except for China, where it is projected to grow a bit slower than the global average.

The expansion of the construction sector leads to a corresponding increase in construction materials use[4], projected in Figure 4.5. As a direct outcome of the projected growth boom in India and many Sub-Saharan African countries, these countries are likely to see the largest growth in construction materials use. The biggest consumer of construction materials in 2017 and 2060 is China, driven by its sheer size, with a growth rate that is projected to reduce but remain sizable. According to the central baseline scenario, Chinese consumption of construction materials is projected to increase only slightly from 22 Gt in 2017 to 24 gigatonnes (Gt) per year after 2025 and then roughly stabilise at around 23 Gt until 2060. Thus, the central baseline scenario foresees a stabilisation (saturation) at least of materials use levels, if perhaps not stocks. Section 4.4.1 discusses the saturation of materials use and its link with economic development in general, while Section 4.4.2 dives deeper into the plausibility of the projections for China, looking in more detail at the uncertainty surrounding the future demand for construction. In the Middle Eastern and African country groups, the need for construction materials is projected to increase from 4 Gt in 2017 to 15 Gt in 2060, while for the non-OECD Asian economies (excluding China) the projected levels are 6 Gt in 2017 and 22 Gt in 2060.

Figure 4.3. In most countries investment is projected to increase over time

Gross investment in tln USD (2011 PPP)

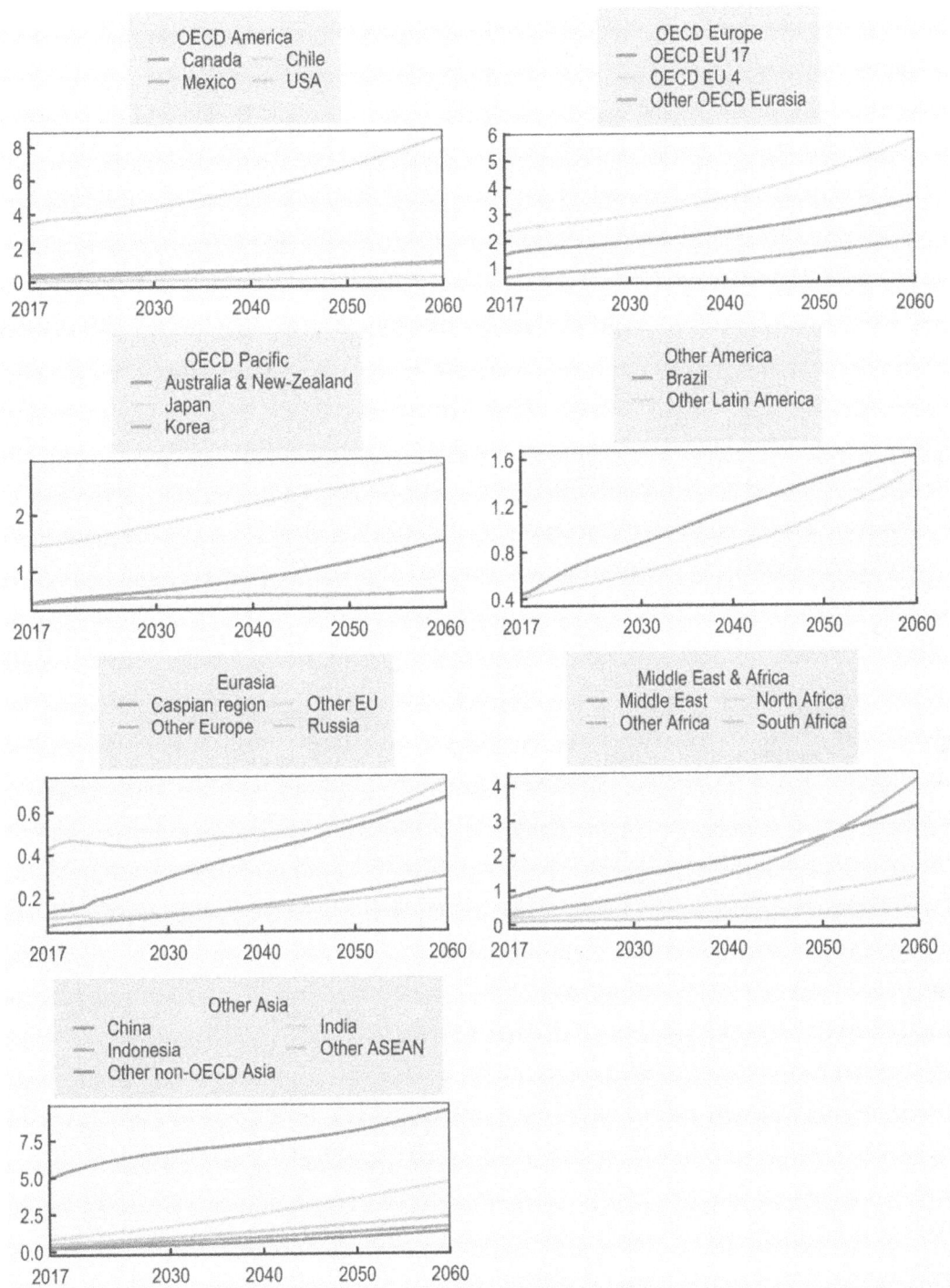

Source: OECD ENV-Linkages model.

StatLink https://doi.org/10.1787/888933884707

Figure 4.4. Construction activity is linked to investment booms

Gross output of the construction sector in mln USD (2011 PPP)

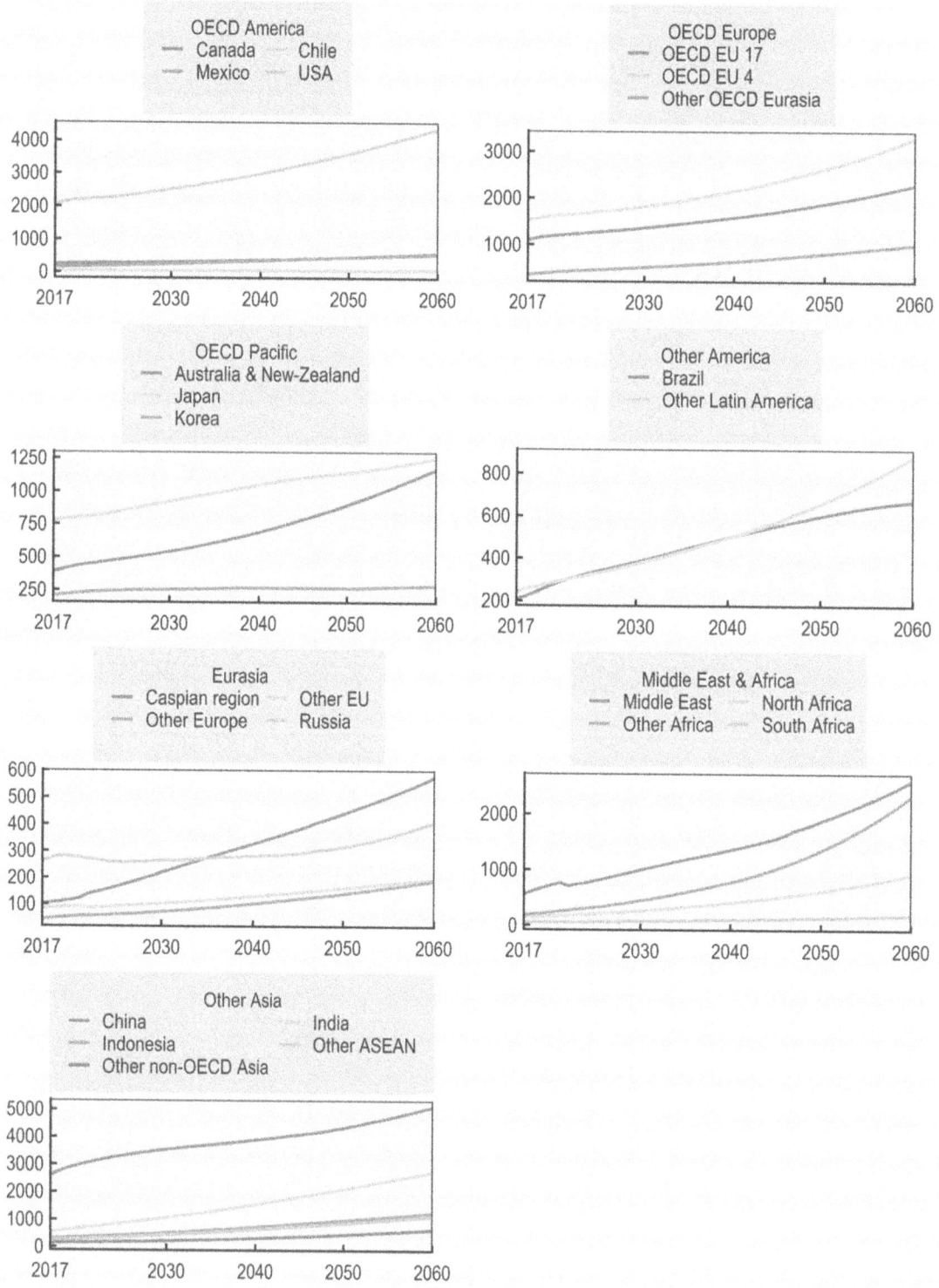

Source: OECD ENV-Linkages model.

StatLink https://doi.org/10.1787/888933884726

Figure 4.5. Non-OECD Asia and Africa are projected to see the strongest growth of construction materials use

Minerals use in the construction sector in Gt

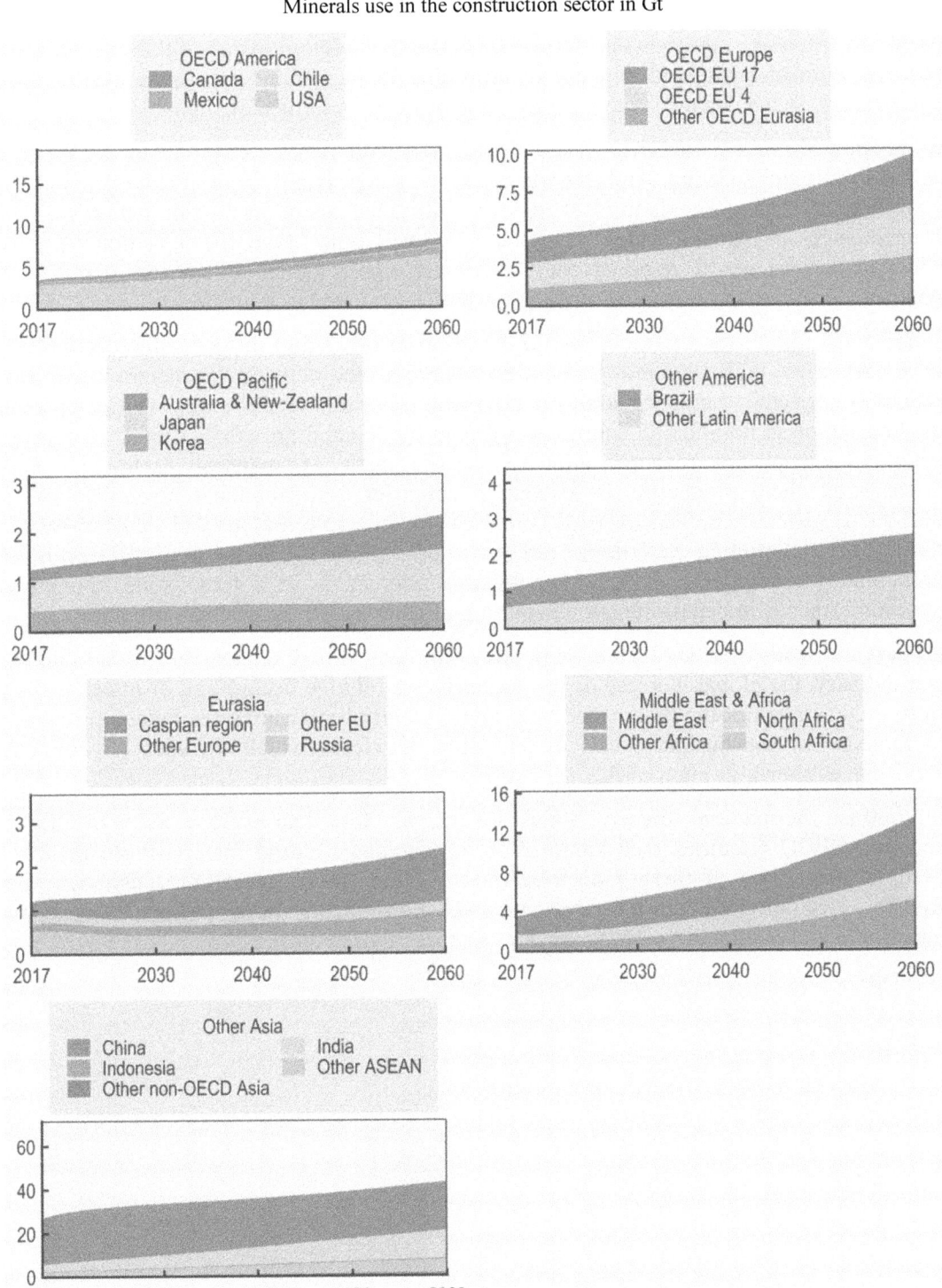

Source: OECD ENV-Linkages model.

StatLink https://doi.org/10.1787/888933884745

4.2. The global rise of services helps reduce materials use

Changes in the sectoral composition of economic activity, resulting from the evolution of demand patterns, affect materials use. As discussed in Section 2.2, the main change is the projected increase in the demand for services. The main determinants of this trend are (i) the evolution of household demand patterns due to increasing per-capita income and ageing; and (ii) the changes in the composition of firms' intermediate demand.

Production of goods increasingly takes place across borders in global value chains (GVC) as a result of globalisation. And imports constitute a significant share of final demand by government and households in most regions. Thus, changes in production patterns differ from changes in demand patterns (Figure 4.6). Over time, trade specialisation patterns adapt to the changes in regional competitiveness, and regional production of different goods and services are projected not to grow at the same rate as demand.

Trade and specialisation patterns also drive material flows and influence the decoupling of materials use and economic growth. For example, the demand for electronics in the OECD is projected to outpace domestic production. For this sector, part of the increase of the production in non-OECD countries is driven by demand growth in OECD countries. Another example is resource-intensive sectors such as materials extraction and processing, where production is located in countries with abundant resources, while demand is generally spread over all countries: for example, output and demand of the fossil fuel and mining sectors do not grow proportionally at the regional level. In particular, the smaller and more integrated a country is in a trade zone, the more important trade is in determining the evolution of its domestic production structure.

As services tend to have relatively low materials intensity in comparison to agriculture and industry, the global increase of services drives a relative decoupling of materials use from GDP growth at the macroeconomic level. Material intensities are indeed projected to remain low for business services and other services (which includes government services), as shown in Figure 4.7.[5]

In contrast, the material intensive industries increase less than average for both OECD and non-OECD countries (see Annex 5.A for detailed sectoral materials use results). Material intensities in these industries (construction, food, agriculture, fisheries and forestry, electricity and utilities, iron and steel, non-metallic minerals) are an order of magnitude higher than those of the services sectors.

Figure 4.6. Projected shifts in regional demand differs from the projected shifts in regional production

Change over the period 2017-2060 in real demand (blue bars) and in real gross output (grey bars)

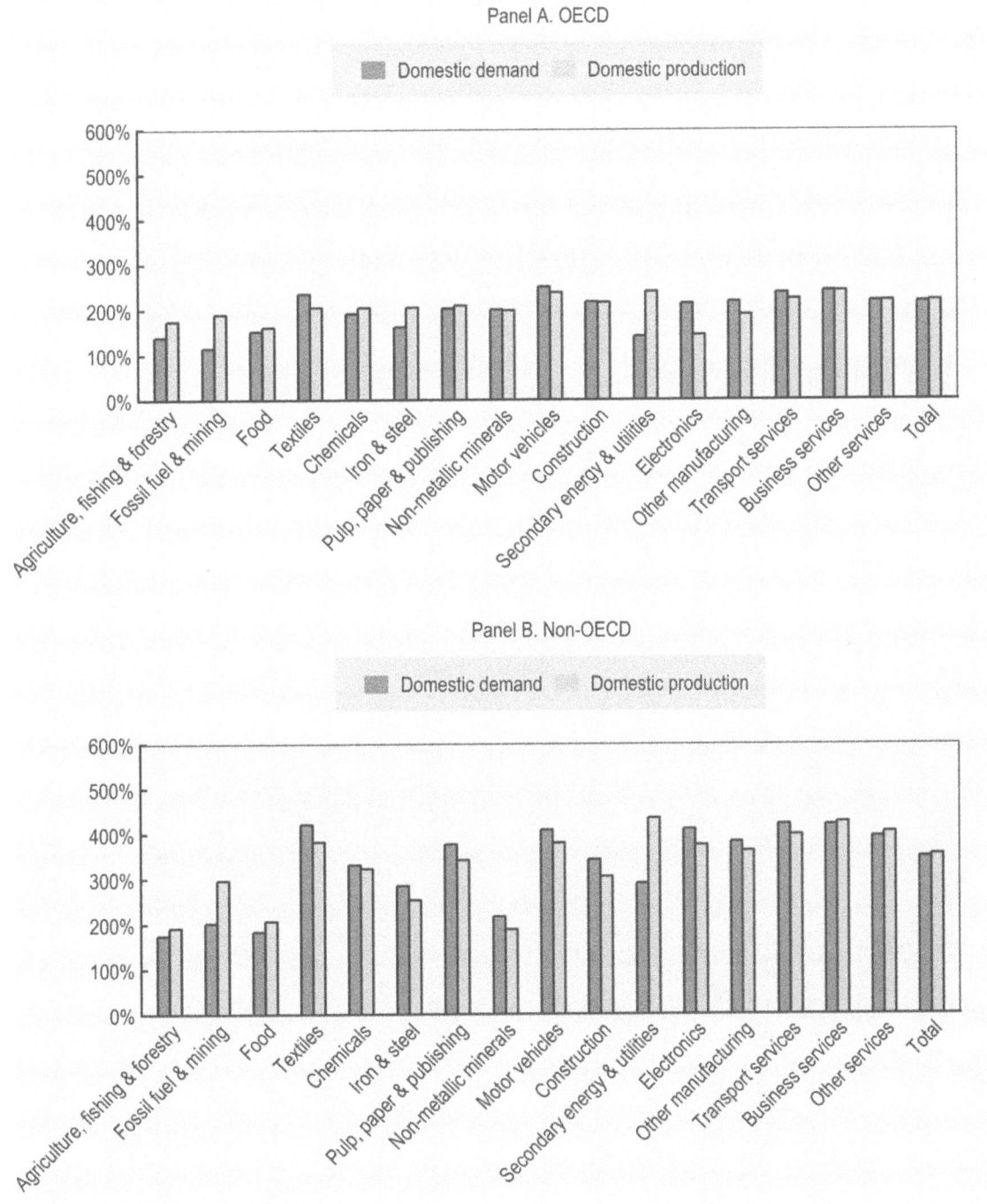

Source: OECD ENV-Linkages model.

StatLink https://doi.org/10.1787/888933884764

Figure 4.7. Less materials-intense services are projected to see above-average growth

Ratio of primary materials use in tonnes over sectoral output in thousand USD.

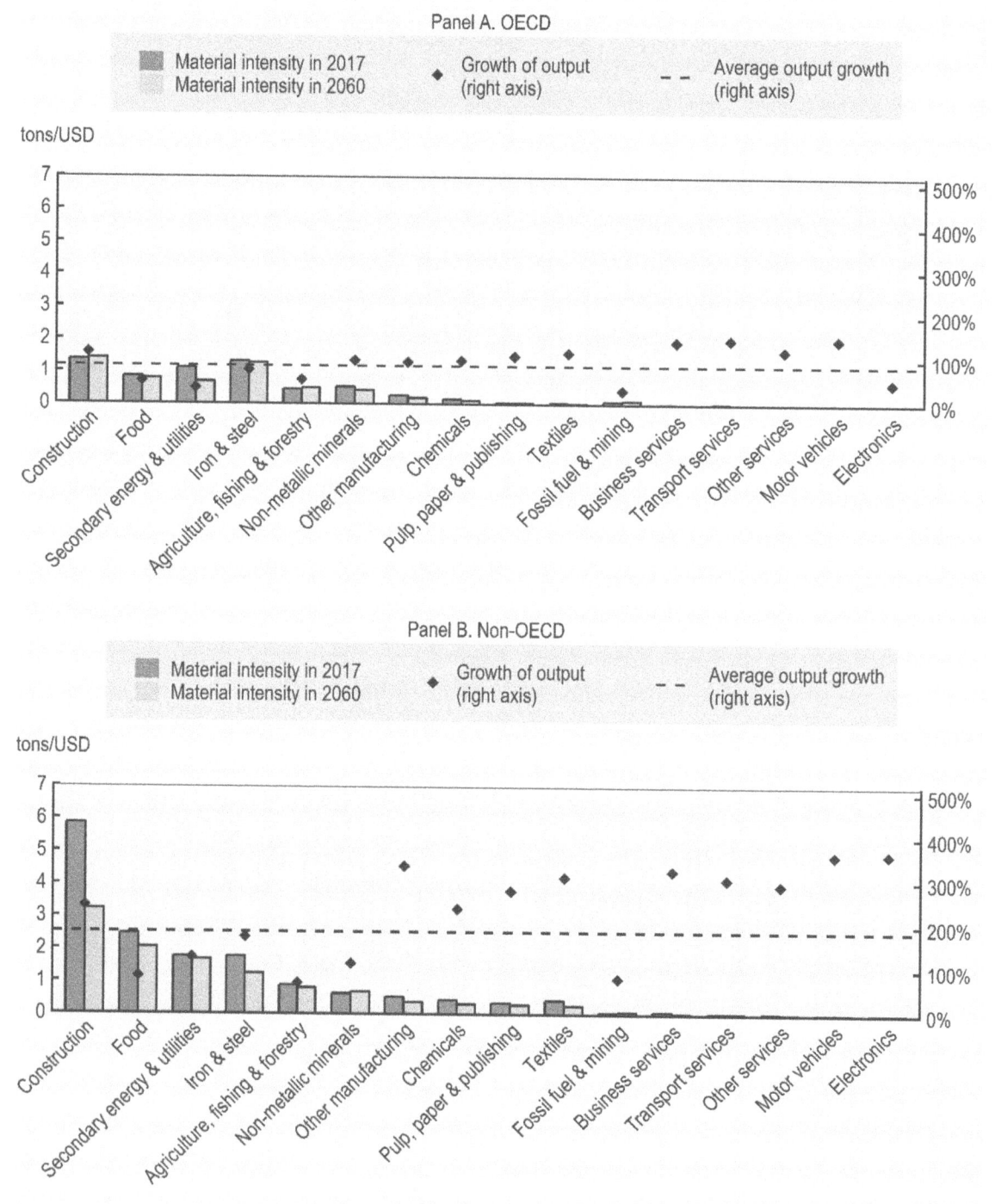

Note: Intensities are in tonnes per one thousand USD. The bars for the sectors with very low values are too small to show on this graph (e.g. Electronics).
Source: OECD ENV-Linkages model.

4.3. There is a gradual transition away from primary material inputs

Changes in production modes influence materials use, as production processes substitute resources, become more efficient and rely on improved technologies. These developments translate into changes in the overall input structure as well as in changes (usually reductions) of unit costs. For example, Section 3.3 shows that in the projection of manufacturing output, unit costs decline over time, and input costs shift away from industrial inputs to services.

Increased efficiency in production processes reduces the materials intensity of production; changing input shares, which are aimed at minimising input costs, can have positive or negative effects on materials intensity, depending on which inputs are cheaper. This can be illustrated once again by the input shares of manufacturing goods, which generally rely heavily on primary materials. As illustrated in Table 4.1, the cost share of inputs of materials-intensive goods, as well as other industrial goods, is projected to decrease over time. This decline is compensated for by the increasing share of services as an input (as shown in Section 3.2). All other things being equal, this implies a decrease in materials intensity in the production of manufacturing goods. This effect is visualised in Figure 4.7 above: the materials intensity is in almost all cases lower in 2060 than in 2017. The size of the effect differs across sectors and regions, and is also affected by the regional aggregation: a shift in production from a more materials-intensive country to a less materials-intensive country will reduce the aggregate materials intensity of the combined region, even if the production technologies at the country level do not change.

Table 4.1. Projected input composition for the production of manufacturing goods

Share of components in total production costs

	OECD			Non-OECD		
	2017	2030	2060	2017	2030	2060
Materials-intensive goods	16%	17%	15%	25%	25%	23%
Other industrial goods	35%	33%	28%	38%	37%	34%
Services	19%	21%	27%	14%	15%	21%
Value added (capital, labour and natural resources)	30%	30%	29%	24%	23%	23%

Note: Materials-intensive goods include agricultural production, forestry, mining (fossil fuels, metals and non-metallic minerals), transformation sector such as metal processing (primary) and reprocessing (secondary), non-metallic minerals transformation and construction.
Source: OECD ENV-Linkages model.

StatLink https://doi.org/10.1787/888933885809

One example where efficiency improvements play a key role is for fossil fuels, as explored in Box 4.1. The box clearly shows how sectoral and technological trends together bring about a relative decoupling of fossil fuel use from GDP growth.

The evolution of production processes and the increasing availability of new technologies drive changes in sectoral and regional productivity. These changes in turn have consequences for competitiveness and thus for production and trade patterns. Furthermore, the availability of recycled (secondary) materials also changes the demand for extracted (primary) materials. This is particularly the case for metals, whose use depends on the regional and sectoral composition of production, the production cost (which differs significantly between primary and secondary production) and the availability of ores and recycled scrap.

Box 4.1. Energy efficiency and fossil fuel use

In the past decades, the production and use of energy has been characterized by significant efficiency improvements. For instance, between 1990 and 2015, while GDP grew by 135%, final energy demand only grew by 50%, while fossil fuel use grew by 60%. Many recent policies also explicitly target energy efficiency. Furthermore, structural trends in the energy sector, not least the rise of renewable energy sources and the electrification of the energy system, affect the demand for fossil fuels. Together, energy efficiency improvements and structural changes mean that fossil fuel use has been growing more slowly than GDP at the global level. In the central baseline scenario, efficiency improvements in energy production and use are projected to occur at multiple levels along the value chain (Figure 4.8). Both supply and demand for energy will be affected.

On the supply side, the reduced use of fossil fuels in electricity is an example of decoupling. Power plants become more efficient, and as a consequence, electricity output grows faster than fuel inputs. Moreover, fossil electricity – which is an important part of total fuel use, especially for coal and gas – is projected to grow less rapidly than total electricity demand, due to the increased uptake of renewables. Furthermore, the difference in growth rates between electricity and final energy demand clearly illustrates the projected electrification of the energy system.

On the demand side, final energy demand is projected to decouple from GDP due to the structural shift towards less energy intensive sectors as well as energy efficiency improvements in equipment (e.g. cars, household appliances, heating) and industrial processes.

Figure 4.8. Energy efficiency, renewables and electrification partially decouple fossil fuel use from GDP

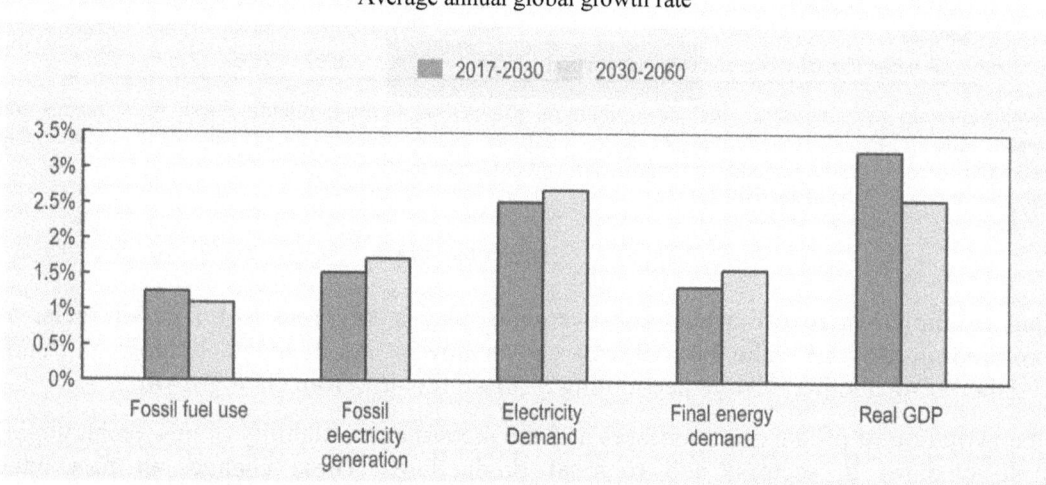

Notes: Fossil fuel use is measured in tonnes, energy in million tonnes of oil equivalent (Mtoe) and real GDP in USD 2011 PPP.
Source: OECD ENV-Linkages model.

StatLink https://doi.org/10.1787/888933884802

Metals are used throughout the economy, and their use is thus influenced by sectoral changes in economic activity. Some of them also show the highest recycling rates of all materials (see Section 6.1). The projected changes in metals use as a result of these technological developments are further explored in this section. The demand for metals is projected to increase following the evolution of regional and sectoral economic activity in the coming decades. Table 4.2 illustrates this by presenting the evolution of the share of different sectors in total demand for ferrous and non-ferrous metals.[6]

Table 4.2. Structural changes affect the demand structure for metals

Sectoral demand for metals: evolution (index 1 in 2017) and shares (percentages)

Panel A. Ferrous metals

	OECD			Non-OECD		
	Growth 2017-2060	Share in 2017	Share in 2060	Growth 2017-2060	Share in 2017	Share in 2060
Chemicals & Plastics	1.7	1%	1%	2.4	0%	0%
Construction	2.3	9%	11%	2.1	16%	21%
Electronics	1.2	1%	1%	3.8	1%	2%
Fabricated metal products	1.6	18%	16%	2.7	9%	16%
Iron and steel	1.9	37%	38%	2.0	16%	20%
Machinery & equipment	1.5	17%	14%	3.3	10%	20%
Motor vehicles	2.2	7%	8%	3.5	3%	7%
Non-ferrous metals	2.3	2%	2%	2.8	1%	1%
Non-metallic minerals	1.6	0%	0%	1.5	1%	1%
Other	2.0	4%	4%	3.9	4%	9%
Other collective services	2.3	2%	2%	4.0	0%	1%
Transport equipment	1.7	3%	2%	3.9	1%	3%

Panel B. Non-ferrous metals

	OECD			Non-OECD		
	Growth 2017-2060	Share in 2017	Share in 2060	Growth 2017-2060	Share in 2017	Share in 2060
Chemicals & Plastics	1.4	2%	2%	2.0	1%	1%
Construction	2.1	5%	6%	2.6	2%	5%
Electronics	1.0	6%	4%	2.9	2%	5%
Fabricated metal products	1.4	13%	12%	3.1	4%	12%
Iron and steel	1.5	3%	3%	2.7	2%	4%
Machinery & equipment	1.4	15%	12%	3.0	10%	30%
Motor vehicles	1.9	8%	9%	2.7	2%	4%
Non-ferrous metals	1.7	37%	38%	1.9	12%	23%
Non-metallic minerals	1.4	0%	0%	1.5	0%	0%
Other	2.0	7%	9%	3.7	4%	13%
Other collective services	2.1	3%	4%	4.1	0%	1%
Transport equipment	1.6	2%	2%	3.5	0%	2%

Source: OECD ENV-Linkages model.

StatLink https://doi.org/10.1787/888933884821

The activities driving metal demand are projected to shift. For example, in the OECD region, the share of construction and motor vehicles in total ferrous metal demand is projected to increase over time, while it is projected to decrease for machinery and equipment, fabricated metal products and, in the iron and steel sector itself (i.e. own-use of iron and steel).

However, these sectoral shifts in the demand for ferrous metals cannot be seen independently from the shift in regional production from OECD to non-OECD countries. While metal demand increases for all sectors in both OECD and non-OECD, the growth rates tend to be higher in the non-OECD region. The projected reduction of the share of ferrous metals in machinery and equipment in OECD countries is balanced by a projected increase in non-OECD countries. In contrast, the reduction in the share of own-use in the iron and steel sector is global. However, this global decline is only relative: in absolute terms, demand by the sector itself is still increasing, but less rapidly than the demand in other sectors.

The projections also show a shift in production of non-ferrous metals from non-OECD to OECD countries. For instance, the share of electronics in the demand for ferrous and non-ferrous metal products is projected to decrease in OECD countries and remain stable in non-OECD countries. In absolute terms, production is projected to expand, with the balance shifting from OECD (where the growth between 2017 and 2060 in non-ferrous metal demand by the electronics sector is around 5%) to non-OECD countries (where demand almost triples).

While both ferrous and non-ferrous metals are an input in the production of machinery and equipment, non-ferrous metals are much less used in construction. Not surprisingly, the own-use within the sector is also substantially larger than the cross-use, i.e. the use of ferrous metals in non-ferrous metals production, and the use of non-ferrous metals in iron and steel production, are substantially lower than the corresponding own-uses.

In addition to these demand changes for metals by sector and region, there are also substitution effects in the production of the metal processing sectors themselves. Indeed, the mining inputs used in the production processes of metals are projected to decrease in the long term (Figure 4.9).[7] This reflects the decreasing intensity of mining inputs in the production of refined metals due to greater efficiency in production processes. Despite the efficiency improvements at the sectoral and regional level, until 2030 the global materials intensity of non-ferrous metals production is projected to increase, especially in the BRIICS countries (panel B). Within the OECD group of countries and in the rest of the world, the share of mining inputs in total production costs is also projected to decline until 2030. According to these model projections, until 2030 the increased global demand for metals and the shift in production towards regions with more materials-intensive production processes dominate the effects of declining materials intensity at the country level for non-ferrous metals. For ferrous metals, the regional decoupling is also apparent at the global scale.

By 2060, the economic decoupling effect dominates at the global level also for non-ferrous metals. As a consequence, the share of mining inputs into metals production is projected to decrease on average by around 1%-point by 2060. As the output of these sectors is projected to continue to increase (overall almost tripling), this only reflects a relative decoupling and not an absolute reduction in materials use.

Figure 4.9. Metals production is only projected to decouple from mining inputs after 2030

Mining inputs as share of total metals production costs (in percentages)

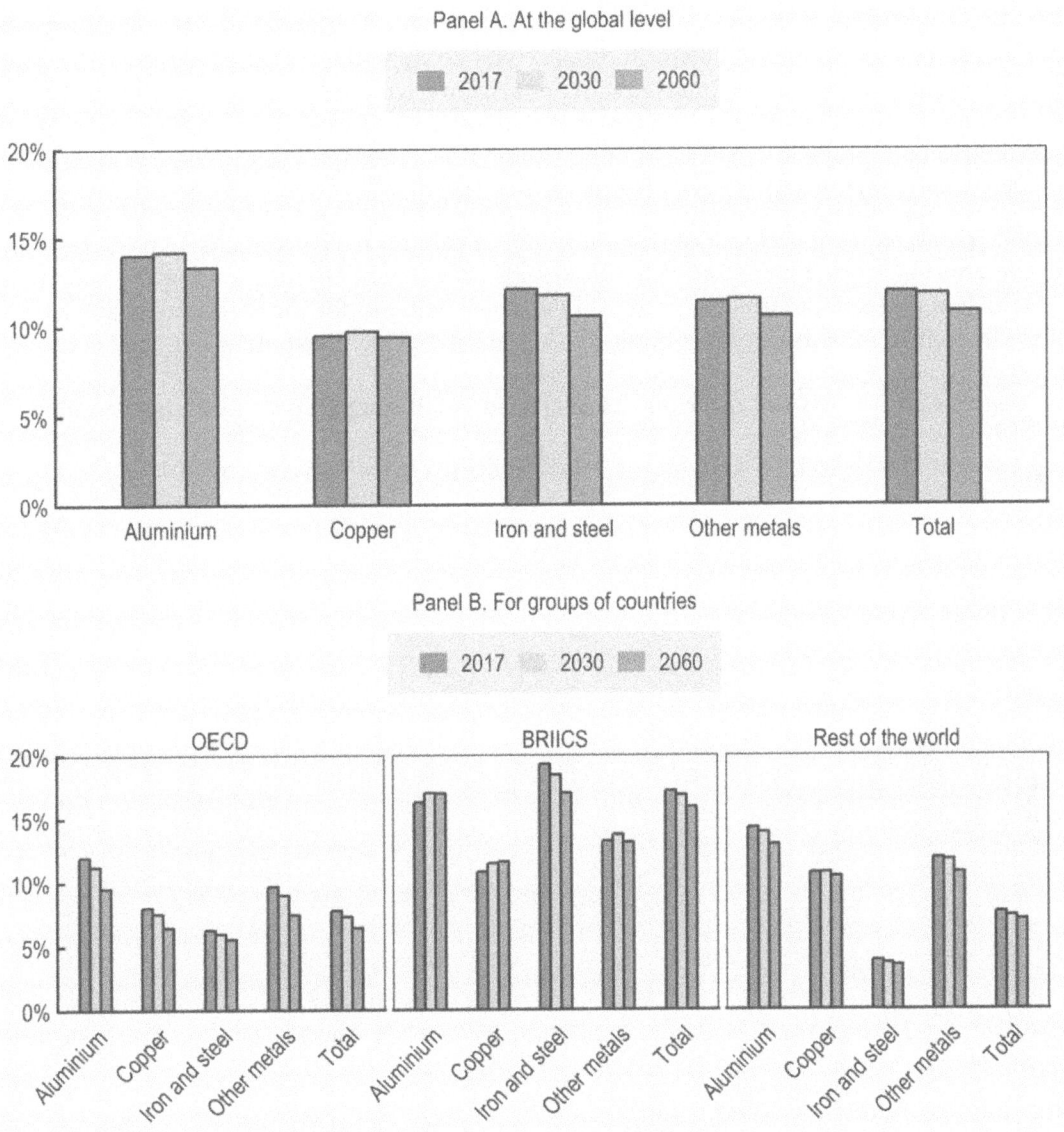

Note: Total production includes primary and secondary material production for each metal, so the decrease also includes the increase in recycling rate.
Source: OECD ENV-Linkages model.

StatLink https://doi.org/10.1787/888933884840

In the central baseline scenario, secondary materials are also projected to grow over time, stimulated by improvements in recycling technologies. As the recycling sector becomes more productive, secondary materials become relatively cheaper than primary materials.[8] In the central baseline scenario without new policies, however, the lower material input costs are outweighed by greater labour costs: secondary metals production tends to be much more labour-intensive than production based on primary ores, especially for non-ferrous metals. As the central baseline scenario projects wages to rise more rapidly than other production inputs, the high labour share put downward pressure on the expansion of the secondary metals production sectors. Chapter 6 describes the dynamics of the competition between primary and secondary materials in detail.

The price dynamics of the recycling sector (for metals and other materials together) are projected to show an improvement in productivity compared to mining (Figure 4.10). The recycling sector provides recycled raw materials for a wide range of sectors: from metals to agriculture, and construction, as detailed in Chapter 6. While the recycling and mining sectors are both projected to expand in the coming decades, the future production costs of recycling are projected to fall well below those of mining.[9]

Furthermore, the production cost (output price) of mining is projected to increase relative to the consumer price index, as mining depends on exhaustible resources, the production of which is projected to involve increasing costs as demand for these resources are scaled up.[10] In contrast, the price of recycling falls compared to the consumer price index.

Figure 4.10. The costs of recycling are projected to fall compared to the costs of mining

Evolution of the output prices of recycling and mining relative to the Consumer Price Index (index 1)

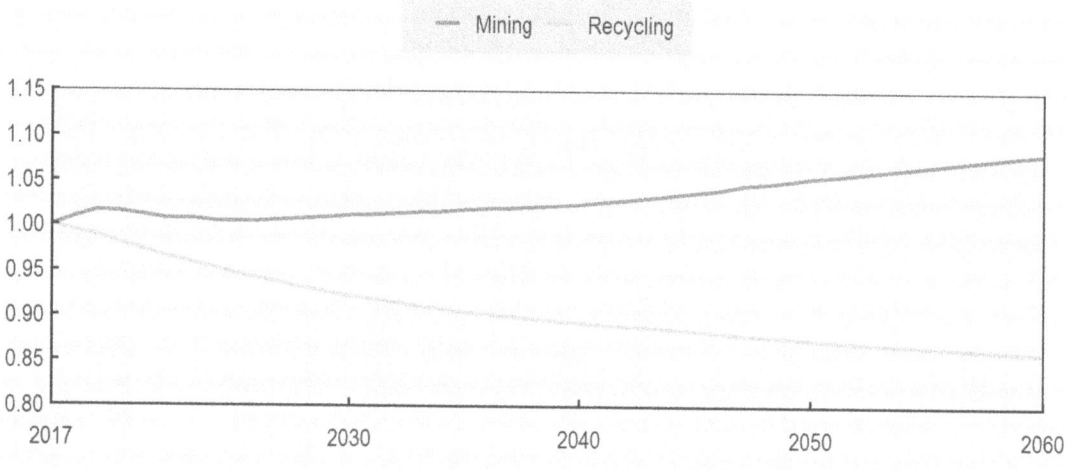

Note: As stated above, mining includes metallic and non-metallic minerals, but excludes fossil fuels extraction.
Source: OECD ENV-Linkages model.

StatLink https://doi.org/10.1787/888933884859

Recycling and mining are, however, only a part of the total production costs of metals processing. Total costs of materials inputs, which comprise mining, recycling and the own-use of the processed metal are in most metal processing sectors less than half the cost of production at the global level (Figure 4.11).[11] The most significant difference between primary and secondary production processes is that primary production is more energy and capital intensive, while secondary is more labour intensive, at least for copper and other non-ferrous metals. This reflects the differences in production technologies. The high labour costs have a dampening effect on the expansion of the secondary sectors.

The material cost share decreases for all metals. In contrast, the shares of labour and other inputs (mostly services) increase overall.[12] These changes are stronger for secondary metals than for primary. The larger labour costs reflect increased wages relative to prices of other inputs. This suggest that the cost structure of secondary metals prevents a significant increase in the share of secondary metal production (see Chapter 6 for details).

Figure 4.11. Secondary non-ferrous metal processing tends to be more labour-intensive than primary

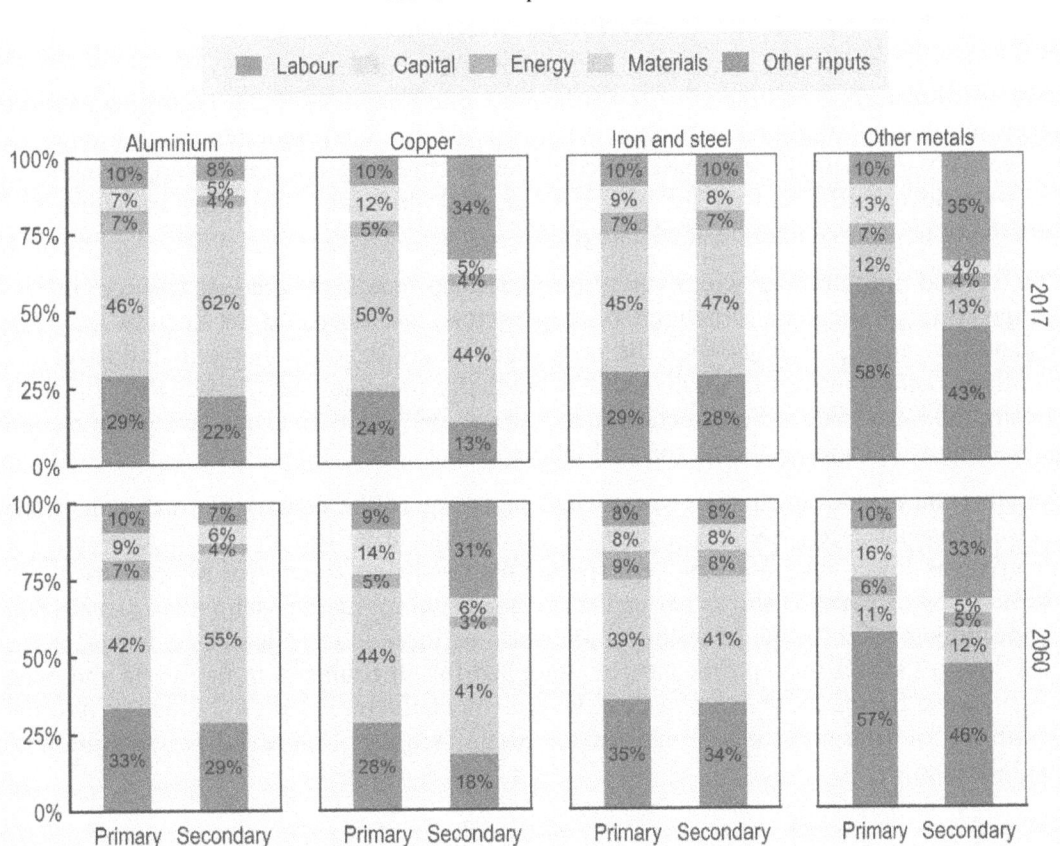

Note: Material input costs comprise inputs of mining, recycling and metals.
Source: OECD ENV-Linkages model.

StatLink https://doi.org/10.1787/888933884878

4.4. Demand for materials declines as economies mature

4.4.1. Economic development may lead to the saturation of materials use

There has been an active debate in the literature on saturation – the stagnation of material stocks per capita as economies mature –, the intensity of materials use over time, and the material environmental Kuznets curve (Bleischwitz et al., 2018[1]; Fishman et al., 2014[2]; Müller, Wang and Duval, 2011[3]; Vehmas, Luukkanen and Kaivo-oja, 2007[4]). All these studies suggest there is an increasing use of materials for infrastructure in early stages of development, which flattens (or even declines) as economic development continues, which is often referred to as saturation. While several authors claim to have found that materials stocks per capita have saturated for some materials in some countries (e.g. Bleischwitz et al. (2018[1]), for steel, copper and cement and Müller, Wang and Duval

(2011₍₃₎) for iron use), the available data on stock accounting is limited and arguably insufficient to robustly identify specific saturation levels for all different materials covered in this report.

Although the absence of stock modelling prevents a direct representation of saturation effects in stocks of materials, the economic drivers of growth affect materials intensity, as discussed above. Thus, these drivers can lead to a stabilisation (saturation) or even decline of demand in the baseline scenario.

The first mechanism that can lead to saturation effects in the model is the geographical shift of global economic activity towards emerging and developing countries and the servitisation of the global economy, as discussed in-depth in Sections 3.1 and 3.2 respectively. This implies increased production in emerging and developing economies vis-à-vis developed countries, as well as a gradual shift away from infrastructure and construction towards other demand categories.

The second mechanism, which is also modelled, entails a gradual change in the composition of the cost structures of key sectors, as discussed in Section 2.3. For instance, in emerging economies the cost share of construction input in infrastructure buildings and housing is projected to gradually decline. Similarly, the input of construction materials in construction reduces over time. In both cases, this phenomenon goes along with increases in the share of labour costs.

Together, these mechanisms imply a gradual saturation of the demand for materials in the central baseline scenario. As a consequence, evidently, the projected demand for materials is lower than a simple extrapolation of existing trends for all countries. An absolute cap on material stocks per capita (as suggested by e.g. Bleischwitz et al. (2018₍₁₎) is not implemented in the modelling framework, as there is insufficient evidence for the peak level of stocks that can robustly be modelled for all countries in the world and even the evidence for the existence of a peak level is inconclusive. Section 7.3 delves into the possibility of saturation of steel stocks per capita (see e.g. Figure 7.3 for saturation of steel demand).

4.4.2. The maturing of China's economy will affect demand for construction materials

The biggest single source of materials use until 2060 projected in the model is construction in China. This is thus an important case to investigate the plausibility of the central baseline scenario on the evolution of materials use and the prospects of saturation of demand (as saturation of stocks cannot be investigated due to lack of data).

The maturing of the Chinese economy is projected to lead to a slowdown of materials use in relation to the growth in GDP. Some decoupling of materials use within the construction sector can be foreseen thanks to increased recycling of some construction materials, as well asincreased efficiency in production. Furthermore, the construction of buildings and infrastructure is subject to saturation effects (see Section 4.4): as housing needs are fulfilled, housing demand stabilises, even as people get wealthier.

As the Chinese economy matures, and especially the expansion of infrastructure slows down, the Chinese construction sector is projected to grow much more slowly than the rest of the economy. While the Chinese economy is still in a boom phase, growth in the construction sector between 2000 and 2016 has not kept up. For example, Figure 4.12 shows that over the last 15 years residential floor space per capita has grown more slowly than GDP per capita: at 60% versus 250%, respectively. Even so, the growth in floor space in China has outpaced that of the EU, Japan and – to a lesser extent – the United States. In 2016, the average area of residential floor space per capita in China is around 35 m², which was already higher than in Japan and very close to EU. In contrast, the USA has a larger residential surface area per capita, reflecting its lower population density than the other three regions.

Figure 4.12. GDP and floor space per capita have moved in the same direction between 2000 and 2016

GDP per capita in thousands of USD (2011 PPP); floor space in m² per capita

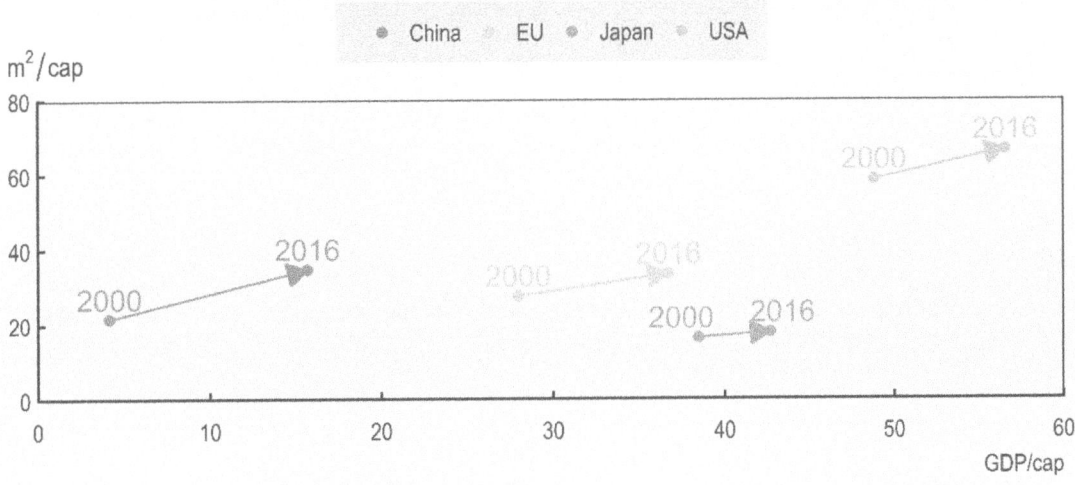

Source: OECD ENV-Linkages model and IEA (2017[5]).

StatLink https://doi.org/10.1787/888933884897

The central baseline scenario of economic growth is highly uncertain, especially when considering a long-term time horizon (see Section 3.4). The uncertainty in economic projections means that there is uncertainty surrounding investments in infrastructure, in the construction sector, and hence in construction materials and potential saturation. In addition to the socioeconomic uncertainties depicted in the alternative baseline scenarios, there are also uncertainties surrounding the evolution of production levels of construction in China specifically.

Figure 4.13 presents construction materials use projections for three alternative scenarios in panel A, and for the alternative socioeconomic scenarios presented in Chapter 2 in panel B. The first scenario in panel A (Existing trends scenario) is an extrapolation of recent historical trends (2000-2015). In this scenario, the construction sector shows a continued increase of production and materials use. There is relative decoupling of construction materials from GDP, but absolute materials use continues to rise steadily. This scenario disregards any effect that changes in the socioeconomic trends over time may have on the construction sector.

The second scenario follows the IEA New Policy Scenario (IEA-NPS scenario) described in the World Energy Outlook (IEA, 2017[5]), which was produced in line with Chinese official projections and in consultation with the Chinese authorities.[13] This scenario only runs to 2040, in line with the time horizon used in the World Energy Outlook. The IEA-NPS scenario assumes that China will adopt a series of policies and measures to adjust the structure of the Chinese economy away from heavy industry (see IEA (2017[5]) for more details). These result in a decrease of the production of the construction sector, and lower materials use.

Figure 4.13. Construction materials use in China is more affected by sectoral assumptions than by the socioeconomic uncertainties

Construction materials use in China in Gt

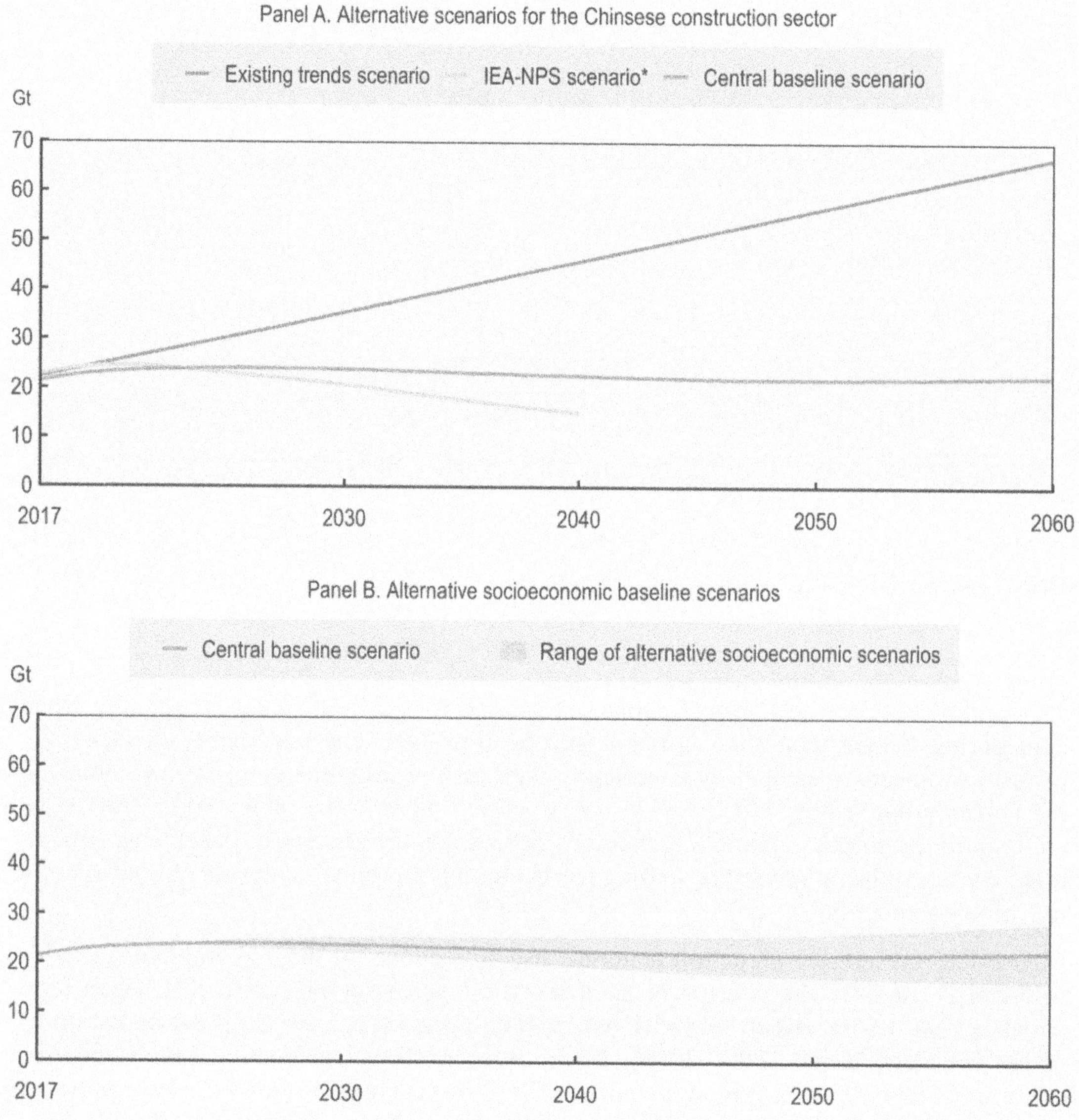

Note: * The IEA-NPS scenario does not extend beyond 2040.
Source: OECD ENV-Linkages model, using IEA data for the IEA-NPS scenario. The *Existing Trends* scenario is an extrapolation of construction output based on historical trends.

StatLink https://doi.org/10.1787/888933884916

The central *baseline scenario* of this report projects a stagnation of construction activity in the coming decades in China, rather than an immediate reduction. Two major assumptions explain the difference between the central baseline scenario and the IEA-NPS scenario. First, the central baseline scenario is built on the assumption that no new policies are implemented in the future, therefore the underlying projections of structural trends (energy as well as construction) are in line with the IEA Current Policy Scenario (CPS) and not the NPS scenario. Secondly, the NPS scenario assumes stronger changes in the sectoral composition of GDP than the CPS, as detailed in Chapter 15 of (IEA, 2017[5]).[14] As a consequence, in the IEA-NPS scenario the economy shifts more strongly away from energy-intensive industries towards services. This structural change goes along with a shift from heavy to light industry, and a reduction of exports. In the *central baseline scenario*, these sectoral shifts also occur but in a smoother and more gradual way.

The *Existing trends* scenario is projected to lead to 56 Gt construction materials use in China by 2050, while the *IEA-NPS* scenario projects a decrease in the use of construction material to 11 Gt. This 45 Gt difference represents more than half of total global non-metallic minerals consumption. The central baseline scenario presented in this report projects the annual use of construction materials in China to peak at 24 Gt per year around 2025, and a gradual decline in materials use thereafter.

The Central baseline scenario thus suggests a saturation of construction material demand that goes beyond the relative decoupling seen in existing trends. But it does not assume an absolute decoupling of construction materials from economic growth – this would require new policies to further stimulate the structural and technical changes needed to achieve the trends of the IEA-NPS scenario.

The alternative socioeconomic assumptions presented in Section 3.4 also influence construction materials use in China, albeit in a more indirect way. Panel B of Figure 4.13 presents the range that is projected across the alternative baseline scenarios. Comparing panels A and B clearly shows that the key assumptions at the sectoral level in panel A have a much larger impact on the projections than the variations in socioeconomic trends in panel B. But panel B does highlight that macroeconomic drivers of materials use can determine whether the demand saturation effect kicks in immediately or is delayed by a decade.

None of these scenarios can *a priori* be disregarded as implausible, as they are building on assumptions that are in principle reasonable. This reflects the uncertainties surrounding these projections. A saturation of the demand for construction materials in China as projected in the central baseline scenario is thus not implausible, but can also not be taken for granted.

Notes

[1] See for instance De Long and Summers (1991[82]).

[2] Other Africa in the figure, which excludes South Africa.

[3] N.B. these are not regional growth rates but shares in global growth.

[4] Sand, gravel and crushed rock, gypsum, limestone, structural clays and ornamental or building stone.

[5] These sectors may indirectly still contribute significantly to materials use through their value

chain, e.g. energy and metals used for office buildings and data centres.

[6] Ferrous metals are essentially iron, steel and the alloys mostly based on iron and steel.

[7] The mining sector includes metallic and non-metallic mineral ores, but excludes fossil fuels extraction (e.g. coal mining).

[8] The recycling sector comprises not only metals, but also all other recycling activities, including e.g. plastics, paper and glass.

[9] The modelling framework can only explain relative prices, not absolute price levels, as the model is homogeneous of degree one in price levels, i.e. a doubling of all prices in the model does not affect the solution of the model equilibrium.

[10] The modelling framework cannot infer whether this translates into an absolute increase in mining prices over time, or a slower decline, as the evolution of the Consumer Price Index cannot be quantified.

[11] Although there are significant differences in labour costs across countries, the baseline evolution over time is shared across regions.

[12] Iron and steel stand out in that analysis since the energy cost share increase significantly, at the expense of the labour share.

[13] The number for this scenario do not match those of (IEA, 2017[25]), as the macroeconomic projections are harmonised with the central baseline projection.

[14] As discussed in (IEA, 2017[25]), the Current Policies Scenario "[...] assumes slower progress in the transition towards a services-oriented economic growth model than in the New Policies Scenario: the share of services in total GDP rises to 60% only by around 2040, which is ten years later than in the New Policies Scenario. Overall, the Current Policies Scenario achieves the same level of economic growth as the New Policies Scenario over the projection period [...]. But the drivers of growth are different between the scenarios, and this has major implications for the structure and shape of the economy in 2040, and for its energy sector."

References

Bleischwitz, R. et al. (2018), "Extrapolation or saturation – Revisiting growth patterns, development stages and decoupling", *Global Environmental Change*, Vol. 48, pp. 86-96, http://dx.doi.org/10.1016/J.GLOENVCHA.2017.11.008. [1]

Fishman, T. et al. (2014), "Accounting for the Material Stock of Nations", *Journal of Industrial Ecology*, http://dx.doi.org/10.1111/jiec.12114. [2]

IEA (2017), *World Energy Outlook 2017*, OECD Publishing, Paris/IEA, Paris, http://dx.doi.org/10.1787/weo-2017-en. [5]

Long, J. and L. Summers (1991), "Equipment Investment and Economic Growth", *The Quarterly Journal of Economics*, Vol. 106/2, p. 445, http://dx.doi.org/10.2307/2937944. [6]

Müller, D., T. Wang and B. Duval (2011), "Patterns of iron use in societal evolution", *Environmental Science and Technology*, http://dx.doi.org/10.1021/es102273t. [3]

Vehmas, J., J. Luukkanen and J. Kaivo-oja (2007), "Linking analyses and environmental Kuznets curves for aggregated material flows in the EU", *Journal of Cleaner Production*, Vol. 15/17, pp. 1662-1673, http://dx.doi.org/10.1016/J.JCLEPRO.2006.08.010. [4]

Annex 4.A. Detailed results and supplementary materials

Figure 4.A.1. The composition of investment expenditure by commodity

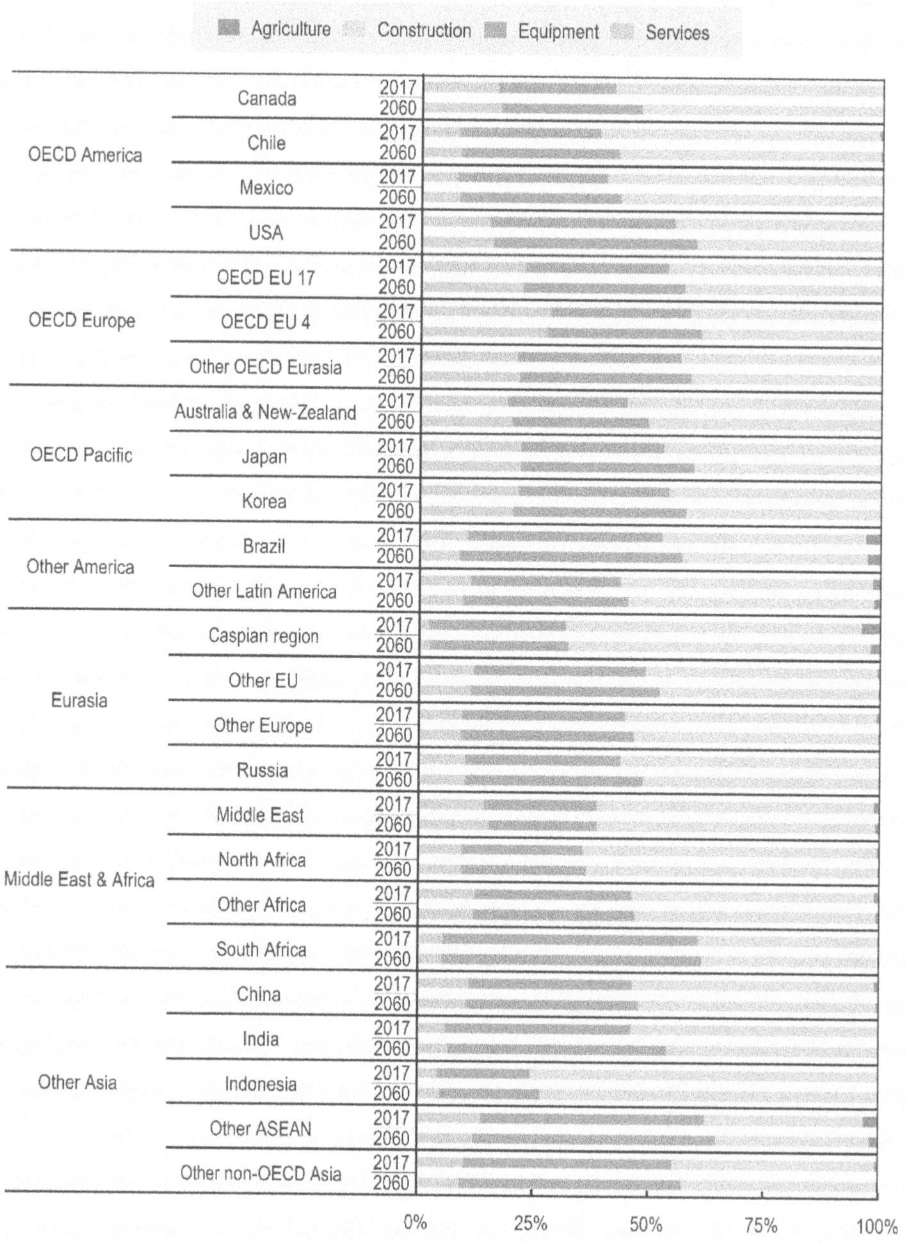

Source: OECD ENV-Linkages model.

Figure 4.A.2. The production of the construction sector by region

Evolution of the production of the construction sector in USD (2011 PPP), index 1 in 2017.

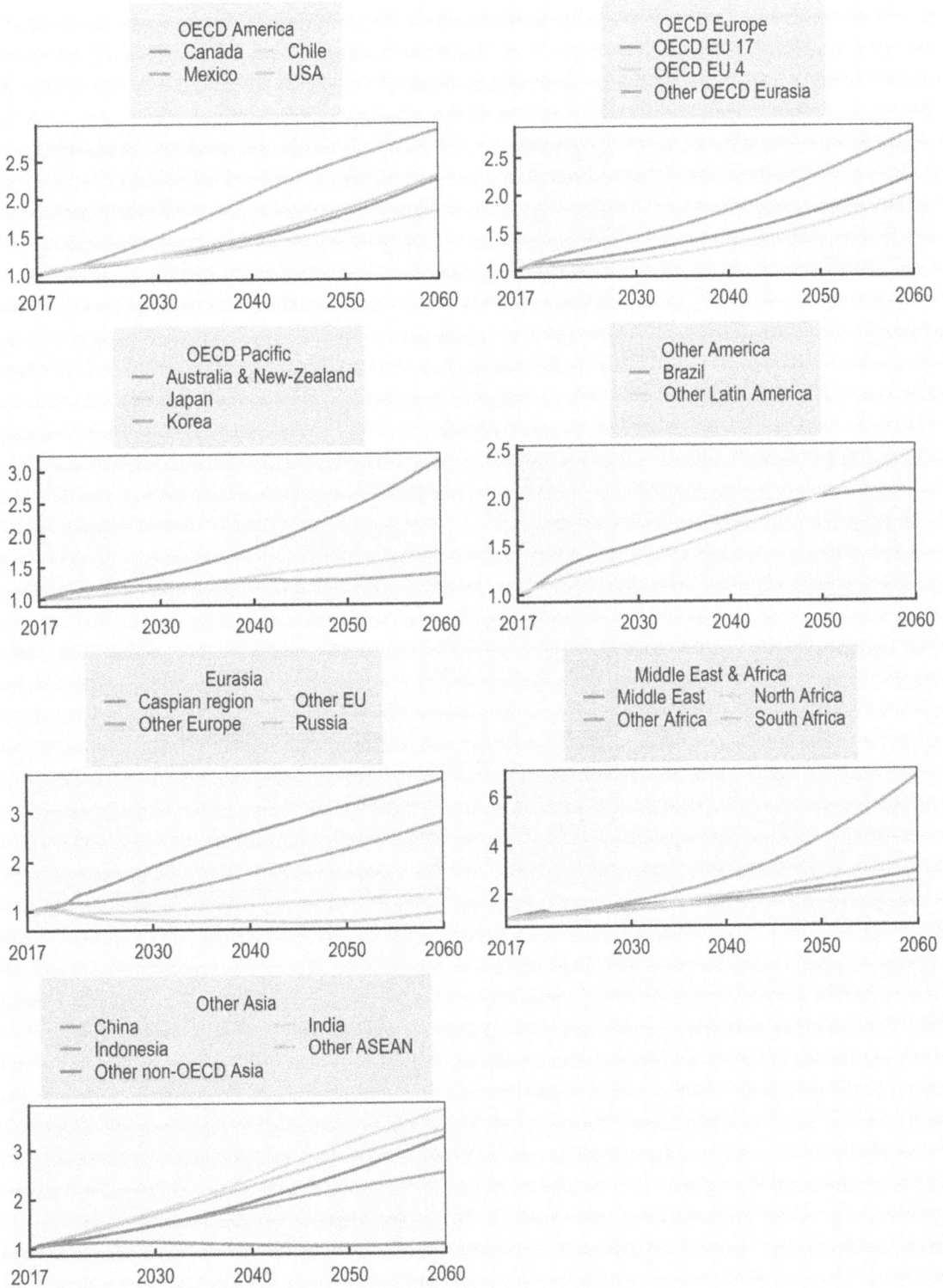

Source: OECD ENV-Linkages model.

Figure 4.A.3. Construction materials intensity of GDP by region

Construction materials intensity of GDP by region measured as the ratio of materials use to real GDP.

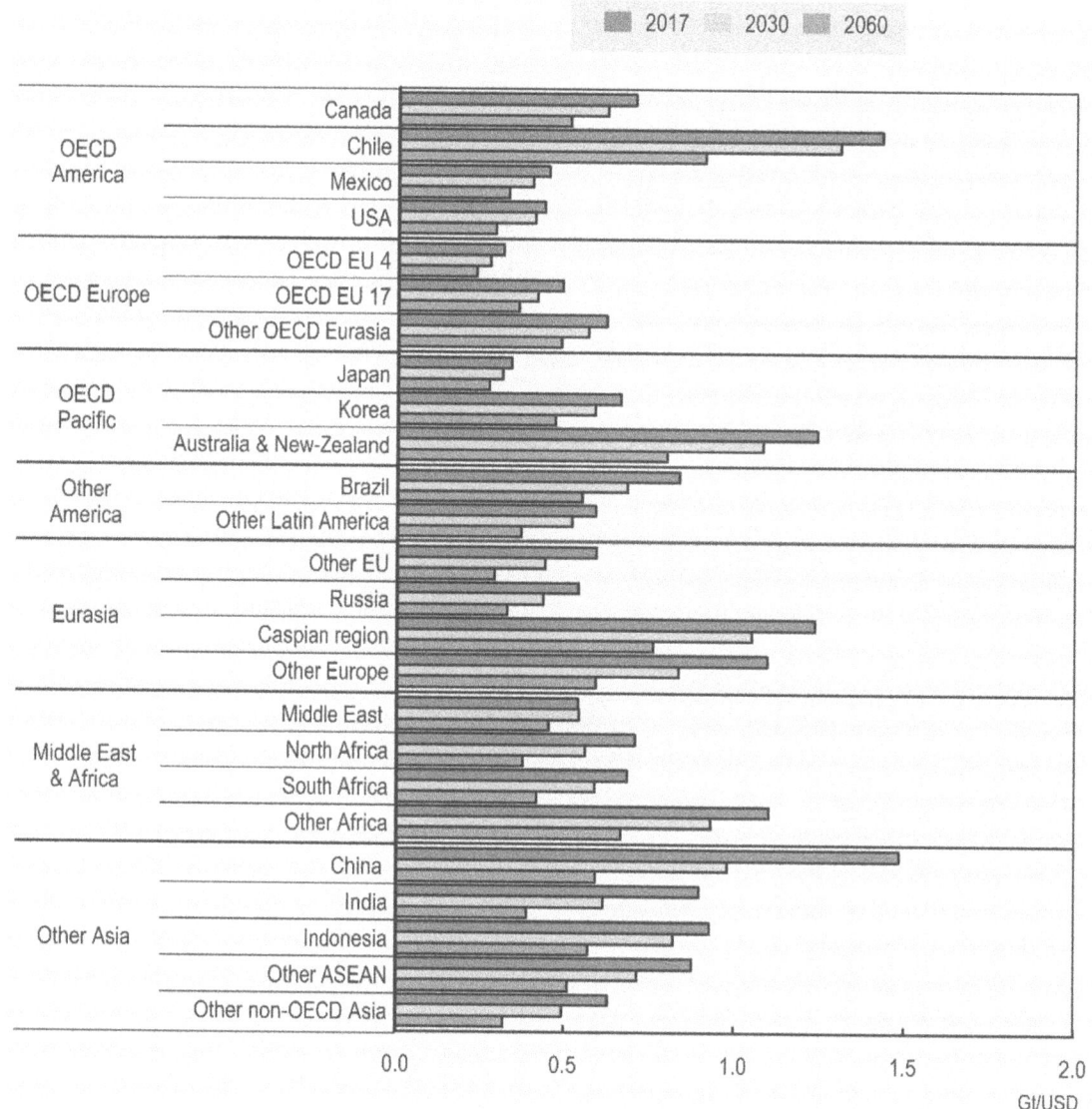

Source: OECD ENV-Linkages model.

Part II.

Materials use to 2060

Chapter 5.

Projections of primary materials use

This chapter presents projections of materials use to 2060 at the global and regional level for four types of materials: biomass, fossil fuels, metals, and non-metallic minerals. These projections focus on three main indicators: primary materials use, materials intensity of GDP and per-capita materials use. The chapter analyses the future dynamics of materials use for these three different aspects, as well as their links to alternative economic and population growth scenarios.

This document, as well as any data and any map included herein, are without prejudice to the status of or sovereignty over any territory, to the delimitation of international frontiers and boundaries and to the name of any territory, city or area.

KEY MESSAGES

To understand the environmental consequences of materials use, and the impacts of policies designed to promote resource efficiency and stimulate the transition to a circular economy, the first step is to take stock of the current use of primary materials (virgin materials sourced from mining and extraction activities). For all countries, non-metallic minerals constitute the bulk of materials used. The use of materials however varies by country, and particularly depends on their development level. Notably, China is a large user of materials given the size of its economy and its need for construction materials to build infrastructure.

Projections and trends

- The use of primary materials is projected to almost double over the coming 5 decades in the central baseline scenario, rising from 89 gigatonnes (Gt) in 2017 to 167 Gt in 2060. In the absence of new policies to improve resource efficiency and stimulate the transition to a circular economy, materials use is projected to increase for all material categories (i.e. biomass, fossil fuels, metals and non-metallic minerals) and to more than double for most of the 60 modelled materials.

- Non-metallic minerals represent the largest share of total materials use, and are projected to grow from 44 Gt to 86 Gt between 2017 and 2060, with the largest growth in tons for sand, gravel and crushed rock. The growth in the use of construction materials also extends to steel. Metal use is smaller in physical terms, but is projected to grow more rapidly and is associated with large environmental impacts.

- While materials use is projected to grow in all countries, growth is strongest in emerging and developing economies. China is projected to remain the largest consumer, but its rate of materials use will slow down. Other non-OECD countries – such as India, Indonesia, and most countries in Sub-Saharan Africa and non-OECD Asia – are likely to undergo an economic and materials use growth spurt, similar to China's in recent decades. Global materials use per capita is also projected to increase, albeit at a slower pace than overall materials use. The total materials use increase is therefore a combination of increasing scale (population growth) and increasing per-capita materials use.

- Despite its 1.9-fold increase, materials use is likely to decouple from the faster GDP growth in most countries (see Figure 5.1). The fast-growing BRIICS countries in particular are projected to experience a strong decoupling in the coming decades if they manage to further mature and diversify their economies. The economies of many African countries are projected to grow rapidly, making Africa the fastest-growing continent by the middle of the century. However, if this growth continues to rely predominantly on the development of materials-intensive industries, for infrastructure construction in particular, as it has done in other regions in the past, the prospects for decoupling in the coming decades are limited.

Figure 5.1. Materials intensity is projected to decrease by 2060

Materials intensity in Gt / tln USD

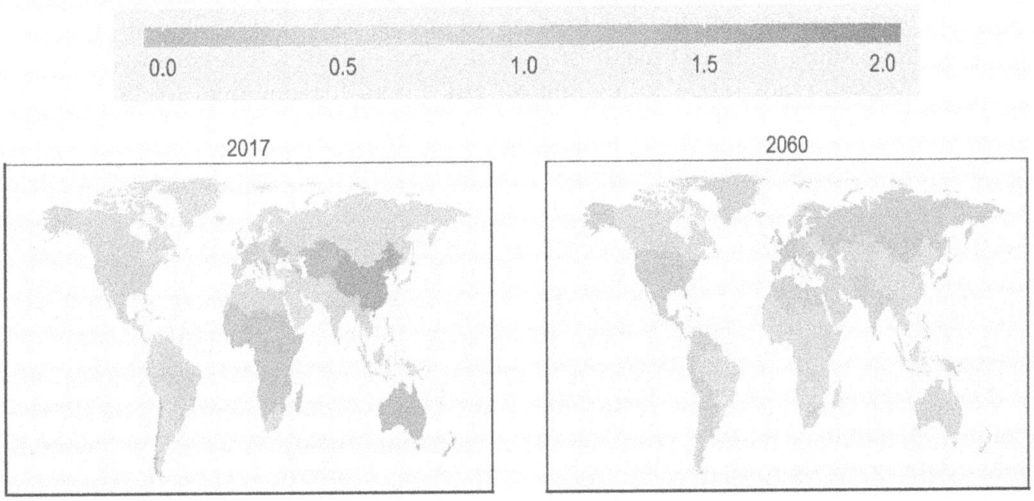

2017　　　　　　　　　　　　　　　2060

Source: OECD ENV-Linkages model.

Areas of uncertainty

The growth in the global use of materials is projected to vary around 20% on each side of the central baseline scenario in 2060. This range considers scenarios with different speeds of convergence and population growth (but identical assumptions on materials productivity). This corresponds to a difference in materials use of -33 to +32 Gt. The largest source of uncertainty stems from the non-OECD Asian countries. These economies currently represent half the world population and could achieve extraordinary growth if they quickly catch up to the living standards of OECD countries. There is much less uncertainty from the socioeconomic drivers surrounding per-capita materials use and materials intensity both at the regional and global level.

Policy implications

This baseline scenario analysis provides several insights for policy making. First, the projected increase in materials use will have environmental consequences that warrant policy action. Furthermore, the projections presented in this chapter help understand the scale of the issue in terms of materials use as well as materials intensity. This debate is often framed as the need to achieving a decoupling between GDP and materials. This chapter helps in providing not only the magnitude of primary materials extraction but also an order of magnitude of the *relative decoupling* that is likely in the absence of new policies to improve resource efficiency and stimulate the transition to a circular economy.

5.1. Development levels affect materials extraction rates

Non-metallic minerals account for the majority of extracted materials, as seen in Figure 5.2 which reveals the 15 most extracted materials globally in terms of weight. Sand, gravel and crushed rock for construction alone represent almost 24% of materials extraction, while other materials used for building and infrastructure construction (e.g. building houses and roads) are also high on the list. Fossil fuels amount to 15% of total extraction. The main extracted metals are iron and copper ores. The top three biomass resources are grazed biomass, straw and other crop residues, which are mainly used as fodder.

Materials extraction varies across regions and development levels (Figure 5.3). This reflects the specialisation of countries: fossil fuels for Russia; biomass for Sub-Saharan Africa (labelled in the figure as "Other Africa", and excluding South Africa); and metals for Australia and New Zealand and Chile. But in most regions, non-metallic minerals are the largest group, given that these consist of relatively low-value bulk commodities (like sand and gravel) that are expensive to import and thus normally sourced domestically. China dwarfs the other regions, with most of its extraction being non-metallic minerals, largely destined for infrastructure.

Figure 5.2. Non-metallic minerals constitute the bulk of materials extraction

Top 15 materials in 2017 in Gt (sorted by global extracted weight)

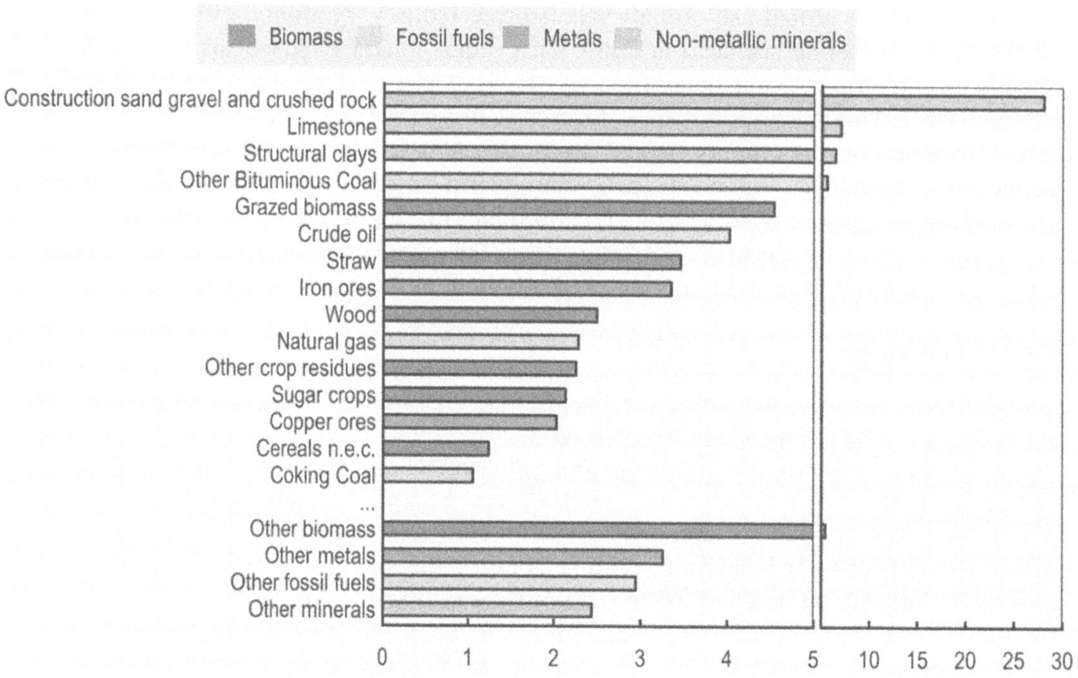

Source: UNEP (2017[1])

CHAPTER 5. PROJECTIONS OF PRIMARY MATERIALS USE | **121**

Figure 5.3. Materials use is heterogeneous across regions and development levels

Materials use by region and material type in 2017 in Gt

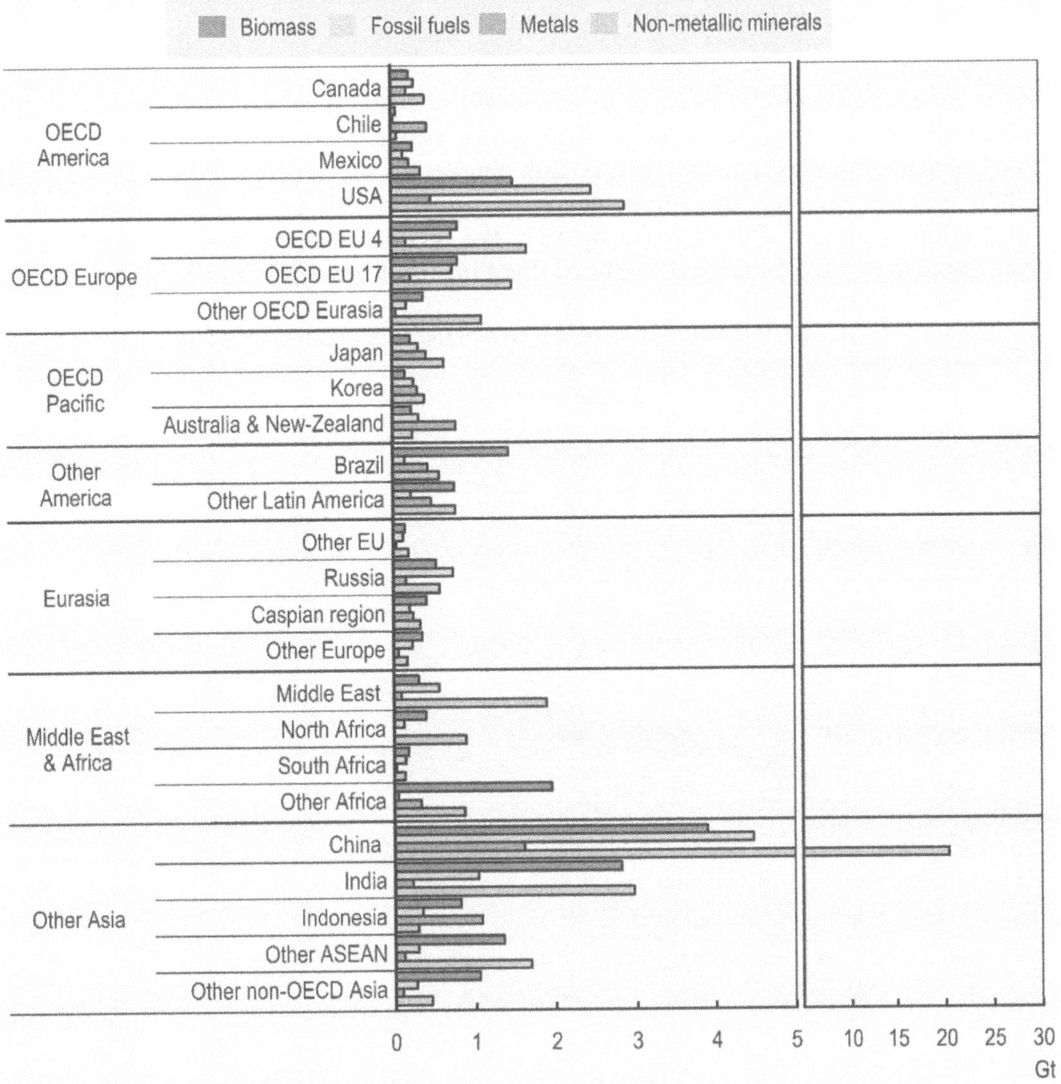

Note: See Table 2.1 in Chapter 2 for regional definitions. In particular, OECD EU 4 includes France, Germany, Italy and the United Kingdom. OECD EU 17 includes the other 17 OECD EU member states. Other OECD Eurasia includes the EFTA countries as well as Israel and Turkey. Other EU includes EU member states that are not OECD members. Other Europe includes non-OECD, non-EU European countries excluding Russia. Other Africa includes all of Sub-Saharan Africa excluding South Africa. Other non-OECD Asia includes non-OECD Asian countries excluding China, India, ASEAN and Caspian countries.
Source: OECD ENV-Linkages model.

StatLink https://doi.org/10.1787/888933884973

5.2. Materials extraction is projected to almost double by 2060

Creating projections of future materials use involves making assumptions about the dynamics that drive changes in the material content of economic flows over time. As outlined in Chapter 2, biomass and fossil fuel resources can be directly linked to the associated production activities, which are covered in sufficient detail in the model. For metals and non-metallic minerals, the projections of materials use are linked to the input of extractive commodities into a processing sector (see Section 2.2). Thus, by using the sectoral and regional economic projections outlined in Chapter 3 and the links between the economic drivers and materials use laid out in Chapter 4, projections are made for all 60 materials covered in the model.

Despite the role of decoupling, which implies an increase in the productivity of materials over time, the projected increase in economic activity in all regions over the coming decades drives a significant increase in global materials extraction – and thus global materials use, as shown in Figure 5.4. Global materials use is projected to reach 111 Gt in 2030 and 167 Gt in 2060, from 89 Gt in 2017 (Table 5.1).[1] Thus, total materials extraction is projected to increase by 88% (i.e. almost double) over 43 years. Box 5.1 compares the global extraction results of this study with previous studies.

Figure 5.4. Global materials extraction is projected to increase across all material types

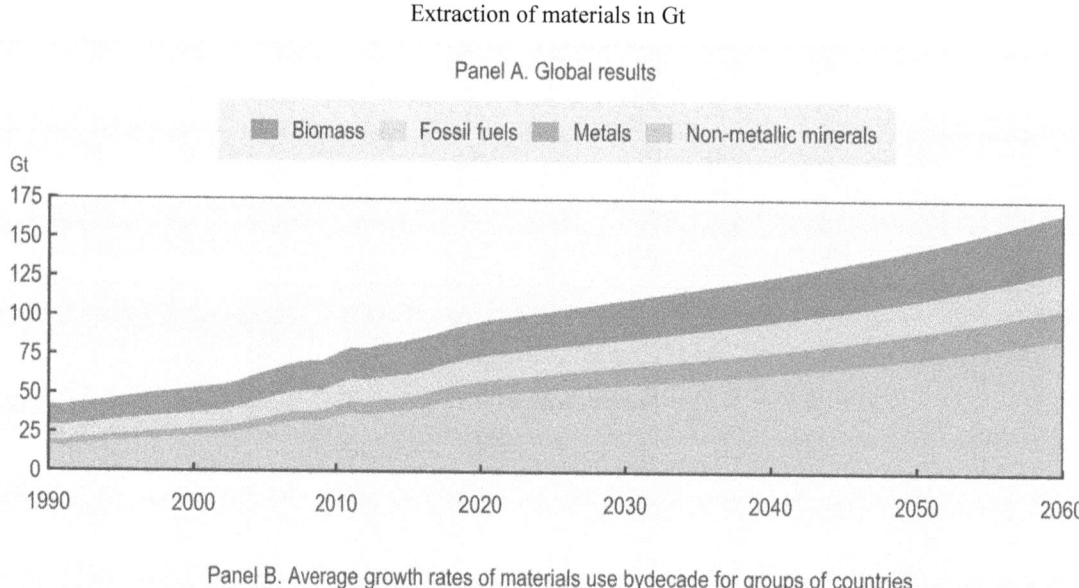

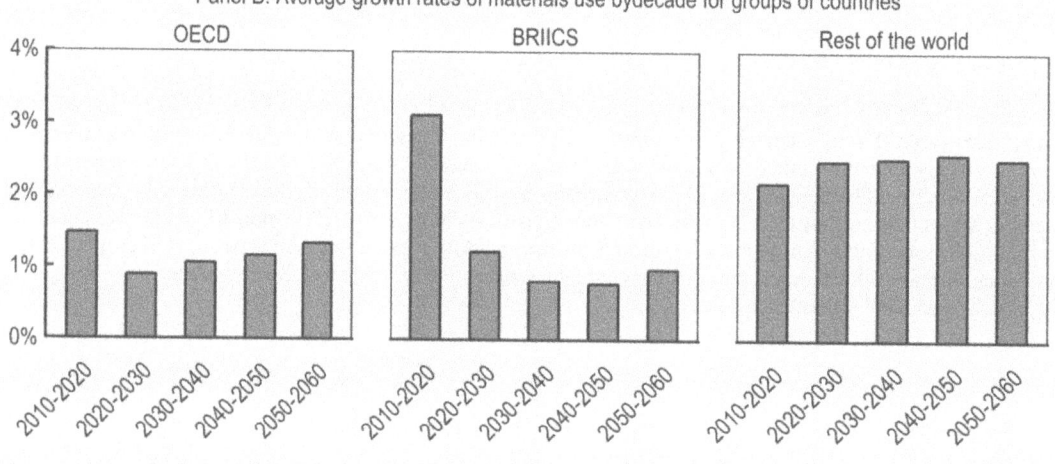

Source: OECD ENV-Linkages model.

StatLink https://doi.org/10.1787/888933884992

Growth rates of materials use at the macro-regional level are strongly affected by the economic dynamics in these regions. They are projected to be fairly stable in OECD countries, in line with the mechanisms explained in Chapter 3 and the notion that these are relatively "mature" economies. In the BRIICS countries (Brazil, Russia, India, Indonesia, China and South Africa), growth rates of materials use are projected to be high in the short run, following the growth boom, and then to fall gradually back to levels similar to those of the OECD. The developing countries (labelled here as rest of the world, ROW) are projected to continue to catch up and sustain growth rates above those of the OECD throughout the projection horizon.

The use of non-metallic materials – mainly *Sand, Gravel and crushed rocks*, as well as *Limestone* and *Structural clays* – is projected to reach 86 Gt in 2060, up from 44 Gt in 2017 (Figure 5.4).[2] This significant increase follows directly from the close link between economic development, investment, infrastructure and construction as highlighted in Section 4.1. As comparing Figure 5.5 with Figure 5.4 shows, materials use in construction is a very large part of non-metallic materials use. The construction sector also uses other materials, not least steel, but these are attributed in the model to the steel processing sector, not to construction (as explained in Chapter 2).

Figure 5.5. Construction materials use is projected to almost double between 2017 and 2060, with the largest growth in sand, gravel and crushed rock

Materials use for materials directly entering construction in Gt

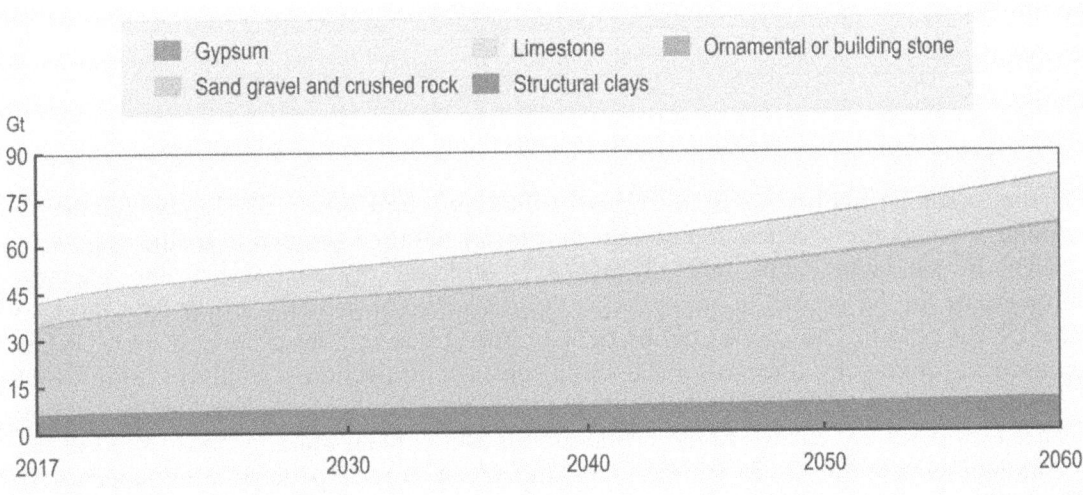

Source: OECD ENV-Linkages model.

StatLink https://doi.org/10.1787/888933885011

The growth of non-metallic minerals and metals has been apparent since the 1990s, driven by China's economic boom, and is projected to continue with growth spurts in other non-OECD countries. The same trend can be identified, at a slower pace, for other materials. A slowdown of the growth rate between 2020 and 2040 for non-metallic minerals is projected, corresponding to a transition phase where China's investment flattens out and the scale of investments in developing economies is still not at full peak, as discussed in Section 4.1.

As shown in Table 5.1, fossil fuel extraction is projected to increase the least between 2017 and 2060, by 63% – as projected in the *Current Policies Scenario* of the World Energy Outlook (IEA, 2017[2]). Biomass extraction is projected to grow slightly more rapidly (+73%), while non-metallic minerals extraction is projected to almost double (+97%). But the fastest growth rate is projected for metals, which increase from 9 Gt in 2017 to a projected 19.5 Gt by 2060 (+126%). For all material groups, these growth rates outpace population growth, but lag behind GDP growth.

Table 5.1. Projections of global materials extraction in the central baseline scenario

		Absolute values				Growth from 2017 levels		
		2017	2020	2030	2060	2020	2030	2060
Population	bln	7.5	7.7	8.5	10.2	3%	13%	36%
GDP	tln USD (2011 PPP)	115	129	175	373	12%	52%	224%
Extraction, *of which*	Gt	89	96	111	167	8%	25%	88%
Biomass	Gt	22	22	26	37	4%	19%	73%
Fossil fuels	Gt	15	16	17	24	7%	18%	63%
Metals	Gt	9	9	12	20	10%	38%	126%
Non-metallic minerals	Gt	44	48	56	86	11%	27%	97%

Source: OECD ENV-Linkages model.

StatLink https://doi.org/10.1787/888933885828

Box 5.1. Global extraction projections in the context of the literature

The most similar publication to this report is the report by UNEP (2017[1]), which uses a similar methodology. When comparing the central baseline projection to the results of UNEP in the table below, one difference is striking: the central baseline scenario projections for the growth in non-metallic minerals are substantially lower than those in the UNEP report. The careful calibration of infrastructure construction in non-OECD countries, driven by the macroeconomic growth projections, imply a significant slowdown in Chinese infrastructure construction. Together with the relative decoupling caused by structural change in the economy this can explain the non-linear trend in the central baseline scenario. In contrast, imagining that growth patterns continue linearly from recent trends (and thus the Chinese infrastructure investment growth continues as today) leads to very similar results as UNEP (2017[1]). An analysis of trends in the economic drivers of biomass use leads to similar results: the central baseline scenario has more profound economic decoupling mechanisms than the UNEP report.

Fossil fuel and metals extraction show very comparable results to those of UNEP (2017[1]). A simple explanation for this is that both models calibrate energy trends to similar detailed sectoral models (i.e. the IEA World Energy Outlook). For metals, the central baseline scenario projects a relatively close coupling between materials use and GDP, and thus results that are closer to those of the UNEP report.

Table 5.2. Projected growth of global materials use in the central baseline scenario is lower than in the 2017 UNEP-IRP report

Materials extraction in Gt

	2050	
	OECD ENV-Linkages (central baseline scenario)	UNEP-IRP (2017[1])
Total	144	184
Non-metallic minerals	73	105
Biomass	33	41
Fossil fuels	21	22
Metals	17	16

Source: OECD ENV-Linkages model, IRP results from UNEP (2017[1]).

The different materials each grow at a specific rate, depending on which economic process they are linked to. Figure 5.6 highlights the evolution of the various materials within each broad materials group. Panel A shows global use in 2017 (as also projected in Figure 5.2) and the additional use between 2017 and 2060; panel B presents this additional use in relative terms, i.e. by normalising the use in 2017 to 1.

Within the group of biomass resources, wood grows more quickly than the other materials. This is because unlike the other biomass materials, wood is related to industrial activities and construction. The food-related materials grow much less rapidly as the demand for food rises slower than overall demand (cf. Chapter 3).

For the fossil fuels, the trends for individual materials follow the detailed energy trends of the IEA (2017[2]). The use of peat decreases significantly in countries that currently use it a lot (incl. many OECD countries), which more than compensates for the increase in use in a number of developing countries. Increases in anthracite use are largely driven by the expansion of the steel sector and roughly match developments in demand for iron ore. The trends for bituminous coal use are driven by expansion of energy use in South-East Asia, as projected by the IEA.

The drivers of increased metal and non-metallic mineral use are more complex, as there are very diverse demand categories for these materials, and widely different patterns of use across countries. The exception is uranium, which follows the nuclear power projection of the IEA (2017[2]).

Figure 5.6. Global use of each material is projected to grow at a specific rate

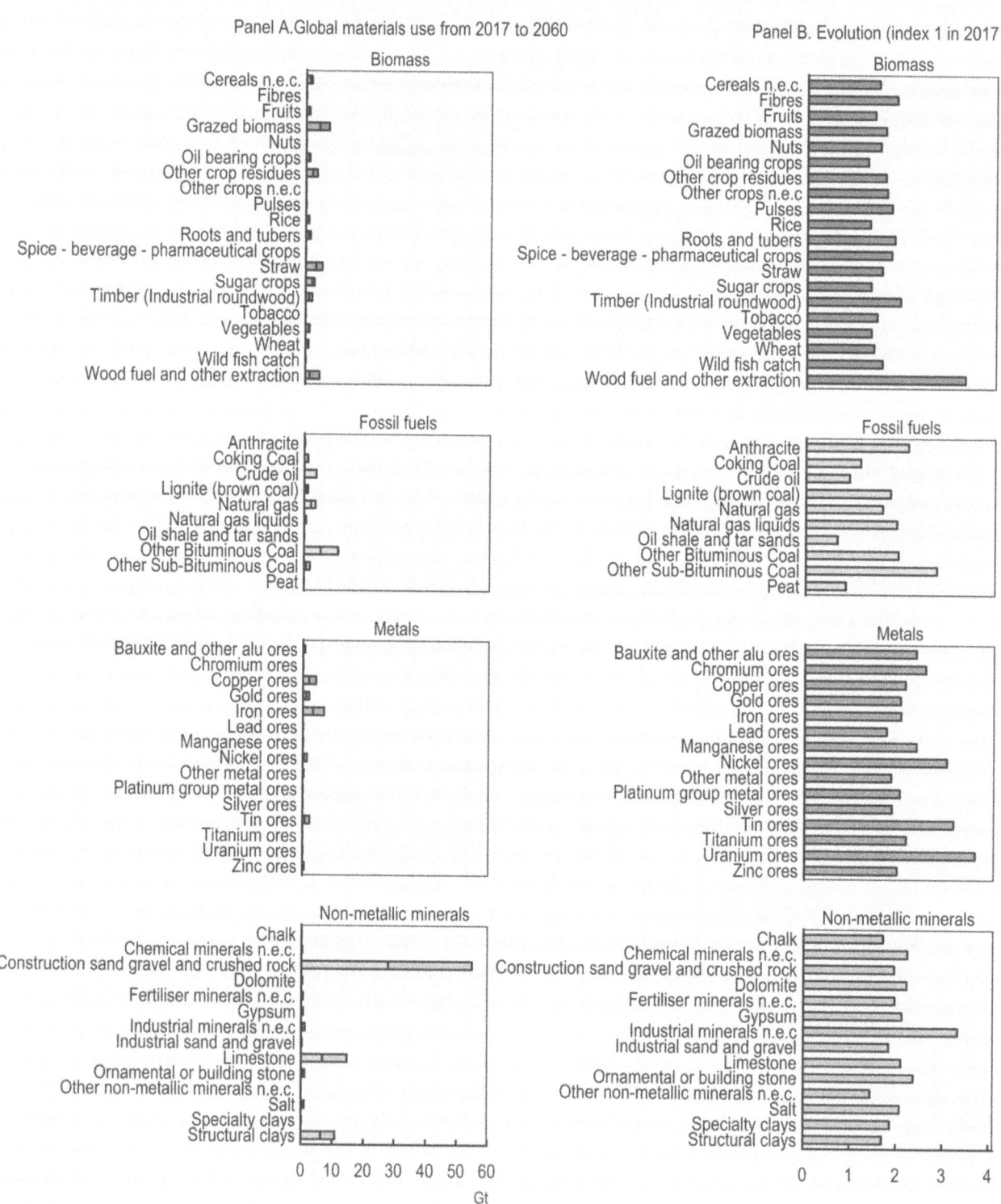

Source: OECD ENV-Linkages model.

Materials use is projected to increase strongly in many non-OECD countries (Figure 5.7). China is projected to remain the main world consumer of materials, but with a slower pace of increase than in recent years. Many other non-OECD countries and regions are projected to undergo a similar economic and materials use growth spurt, including India, Indonesia, Sub-Saharan Africa (Other Africa in the figure) and non-OECD Asian countries. OECD countries are projected to keep increasing their materials use over the period, but at a much slower rate. For some countries such as Russia the model projects declining materials use, mostly because they are projected to follow low economic growth paths. As shown in Figures 3.2 and 3.3 in Chapter 3, population and income growth are projected to be relatively small in Russia, especially in the coming two decades.

Figure 5.7. Materials use is projected to increase in most countries

Materials use in Gt

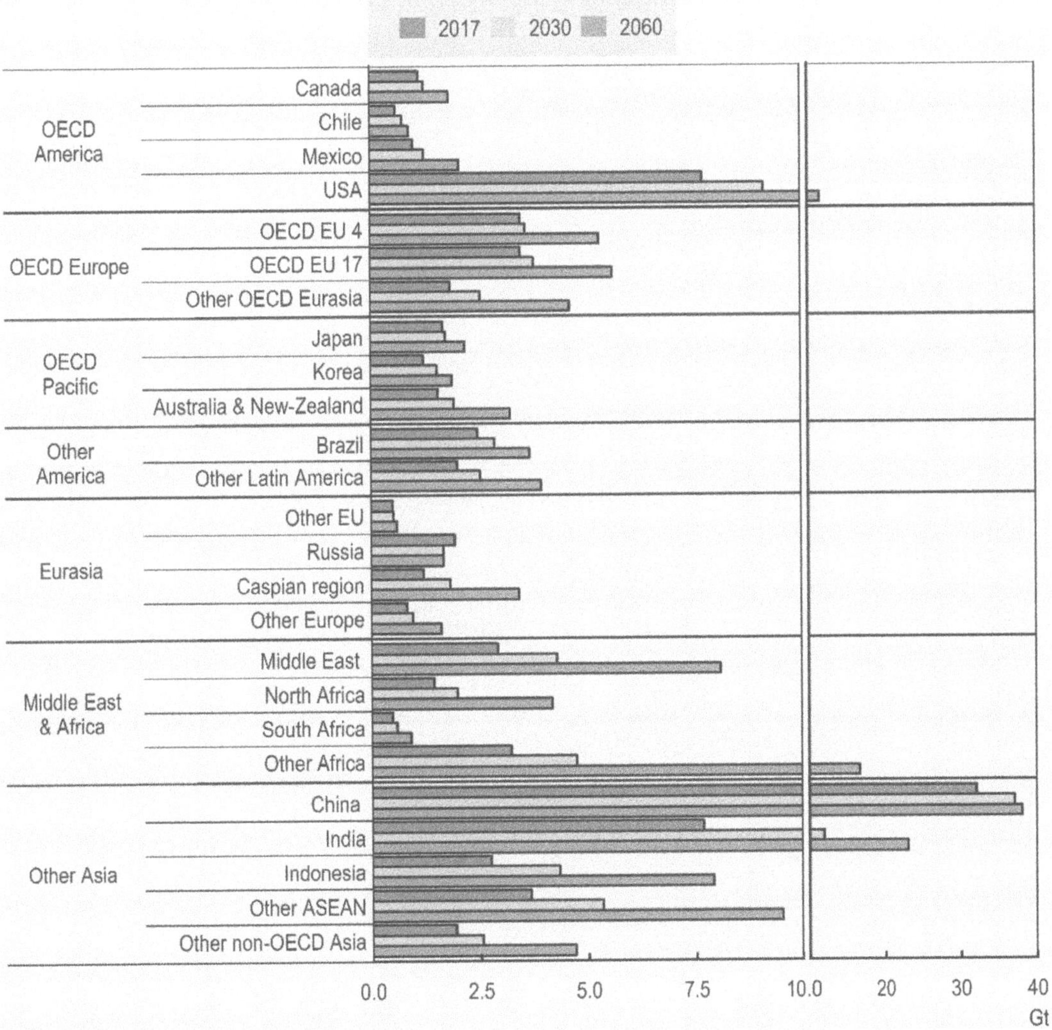

Source: OECD ENV-Linkages model.

StatLink https://doi.org/10.1787/888933885049

5.3. Materials intensity is projected to decline

Comparing the changes in GDP (Figure 3.4 in Chapter 3) and materials extraction (Figure 5.4) shows that while GDP is projected to more than triple, projected global materials extraction not even doubles. Thus although an absolute increase in materials use is projected, the increase is projected to be slower than the growth of GDP, i.e. a relative decoupling is projected. GDP is indeed projected to increase by 224% to 2060 relative to 2017, while projected global materials extraction only increases by 88%. That assessment holds for all four groups of materials, whose growth between 2017 and 2060 vary from 63% for fossil fuels to 126% for metals (see Table 5.1).[3] Panel A of Figure 5.8 presents the materials intensity of GDP at the global level.

Figure 5.8. Global materials intensity is projected to decrease by 1.3% per year on average, and especially strongly in emerging economies

Materials intensity in Gt / tln USD (i.e., kg/USD)

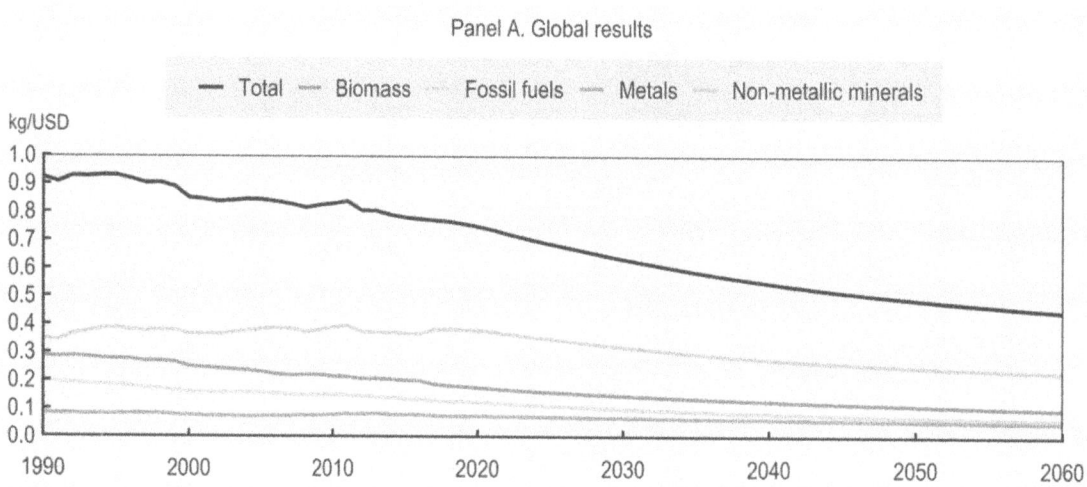

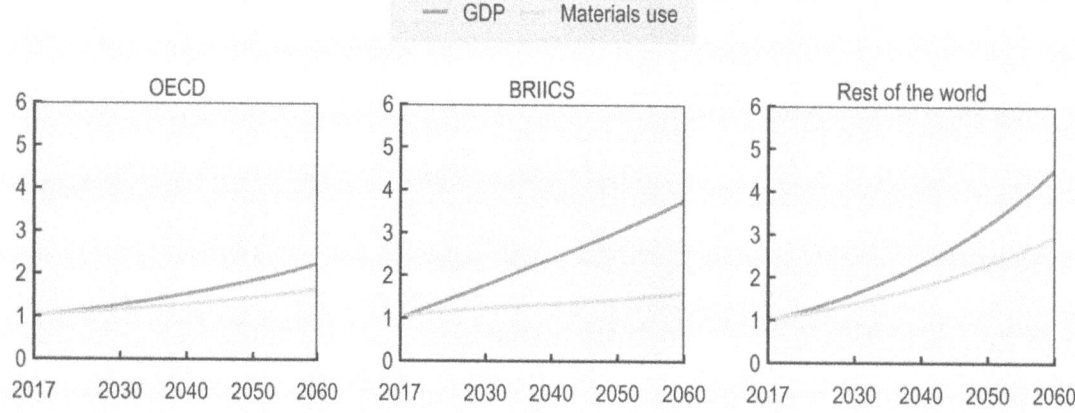

Source: OECD ENV-Linkages model.

Box 5.2. Materials intensity projections in the context of the literature

The variations in the projected annual growth rates of materials intensity reflect the material and time period considered. The projected materials intensity growth rate is around -1.3% per year on average between 2017 and 2060. Thus, the average decoupling rate – measured as the decline in the ratio of total materials use in Gigatonnes to GDP in 2011 purchasing power parity (PPP) exchange rates – equals 1.3%. This indicator is however very sensitive to the metric used to measure GDP. Using market exchange rates to aggregate GDP across regions implies a larger share for OECD countries and thus different decoupling rates (0.9% on average per year between 2017 and 2060). Taking 2005 PPP exchange rates to aggregate across regions – as done in the earlier literature – instead of 2011 values leaves the decoupling rate roughly unchanged.

The projected average decoupling rate (1.3%) is somewhat higher than recent historical trends (0.7%), as shown in the table below. This is not surprising when taking into account the boom in materials use in China over the 1990-2007 period. Decoupling rates for biomass, metals and fossil fuels, not shown in the table, are projected to gradually decrease. A range of factors play a role here, including changes in the sectoral structure of the economy and shifts in regional production towards emerging and developing economies. A comparison with published studies reveals a wide range of decoupling rates. Some studies project rates of up to 2% per year, while others end up with no decoupling. For example, in the UNEP (2017[1]) report, decoupling effects occur at the sectoral and regional level, but are compensated for at the global level by the increasing weight of materials use in relative materials-intensive regions. Since most studies are unclear about the exchange rates used for aggregating across regions, comparison is difficult. This is clearly an under-researched topic that requires more detailed analysis.

Table 5.3. Materials intensity improvements compared with historical rates and other studies

Annual growth rates of materials intensity

	Period	All materials
Historically observed rates – 2011 PPPs	1990-2016	-0.7%
Central baseline scenario – 2011 PPPs	2017-2060	-1.3%
Central baseline scenario – 2011 MERs	2017-2060	-0.9%
Central baseline scenario – 2005 PPPs	2017-2060	-1.3%
Cambridge Econometrics (2014[3])	2014-2030	-0.9%
Böhringer and Rutherford (2015[4])	2020-2050	-2%
Schandl et al. (2016[5])	2010-2050	-1.5%
Hu, Moghayer and Reynès (2015[6])	2010-2050	-2%
UNEP (2017[1])	2015-2050	0.0%

Source: OECD ENV-Linkages model for own results, and other studies detailed in table.

StatLink https://doi.org/10.1787/888933885847

For biomass, fossil fuels and metals, there is a clear downwards trend in the projections until 2060, confirming a relative decoupling of materials use from GDP growth. For non-metallic minerals, the increase noticed in the historical data from 1990 to 2015 is projected to level off in the short run. The intensity of non-metallic minerals use in GDP is projected to decrease after around 2020, driven by the projected slowdown in China's growth. Fast growth of infrastructure in other emerging economies does not prevent the decrease of global non-metallic minerals intensity (see Section 4.1 for a discussion of how development patterns affect non-metallic minerals use). Box 5.2 compares the global materials intensity results of this study with previous studies.

Figure 5.9. Materials intensity is projected to decrease everywhere, but not at the same pace

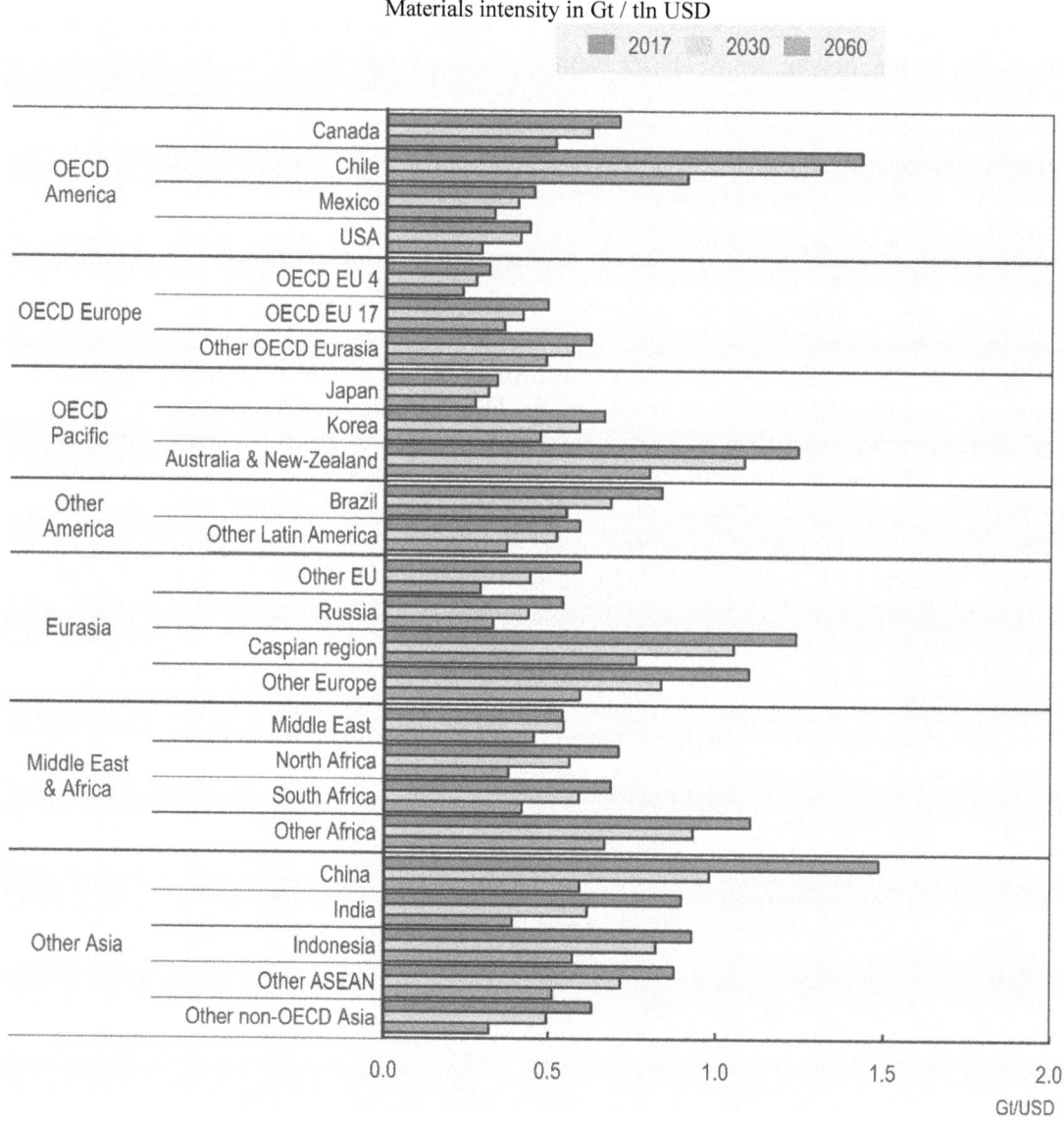

Source: OECD ENV-Linkages model.

StatLink https://doi.org/10.1787/888933885087

Relative decoupling trends are projected to occur in OECD, BRIICS and the rest of the world (panel B of Figure 5.8), but there are some interesting regional differences. Most importantly, the economic dynamics in the projection for the BRIICS region are clearly visible: rapid growth of GDP with a shift in the structure of the economy towards more services combine into a strong decoupling even though overall materials use is projected to grow more than in the OECD. The projected trend in the developing countries (rest of the world) involves the acceleration of both economic activity and materials use, leaving less room for decoupling.

The projected decline in global materials intensity as illustrated in Figure 5.8 is shared by almost all countries (Figure 5.9). The projected reductions are fairly strong in emerging economies, but OECD countries are also projected to have significant reductions in materials intensity. As discussed in Section 3.2, this is to a large part related to the shifts in the sectoral structure of the economy towards less material intensive sectors, not least services, and does not imply a very significant decoupling of materials use and economic activity at the sectoral level.

As mentioned above, the unfavourable demographic developments plus weak income growth projected for Russia imply that the scale of economic growth is projected to be small. But Figure 5.9 shows that the projected decline in materials use in Russia does not only come from a weak economy, but also from a further relative decoupling of materials use from economic activity (mostly driven by structural change).

5.4. Materials use per capita is projected to increase in most countries

The use of materials per person is projected to increase on average (panel A of Figure 5.10), but more slowly than total materials use – an increase by 44% to 2060 – as population growth does not affect this indicator. The main driver of per-capita materials use is income growth (cf. Figure 3.4 in Chapter 3). Structural change (cf. Figure 3.5 in Chapter 3) also significantly affects the projected trends due to varying growth rates in materials-intensive sectors versus sectors that rely less on materials, as discussed in Section 3.2.

Globally, the projected growth of extraction of non-metallic minerals, led by the development of emerging economies, largely explains the doubling of overall materials use over the next 50 years. Compared to 2017, in 2060 per-capita use is projected to increase by 20% for fossil fuels, by 27% for biomass, by 45% for non-metallic minerals and by 66% for metals. Overall, material use per capita increases by 44% on average while GDP per capita increases by 138%. Comparing these indicators emphasizes the significant effect of regional and sectoral shifts. For example, the projections suggest that while total materials use keeps increasing between 2030 and 2040, per-capita use remains flatter, indicating that in this period the economic drivers of materials use roughly cancel each other out, and the increased scale of total materials use is mostly driven by population growth.

Materials use per capita is projected to increase in most countries, reflecting structural shifts and rising income levels. The projections of regional materials use per capita are presented for the macro regions in panel B of Figure 5.10 and in full regional detail in Figure 5.11. As total economic activity and materials use roughly scale with population, it is not surprising, that per-capita use is projected to be much more equal across countries than total materials use.[4]

Some resource-rich countries are, however, projected to see a stabilisation – and in the case of Russia even a reduction – in materials use per capita between 2017 and 2030 (Figure 5.11). Given that in many cases these economies have among the highest initial levels of materials use per capita, and that projected materials use per capita increases rapidly in the emerging economies, this reflects a convergence of materials use patterns across countries.

When looking in more detail at the largest materials category, construction materials, their use per capita is projected to grow for OECD and emerging and developing economies alike (Figure 5.12). Growth is generally projected to accelerate after 2030, except in China where the projected slowdown in economic growth also dampens growth in materials use per capita. Despite the relative decoupling found for overall materials use, many countries are projected to reach a use level of 10 tonnes per capita. Given the dominance of construction materials in overall materials use, this result is not surprising.

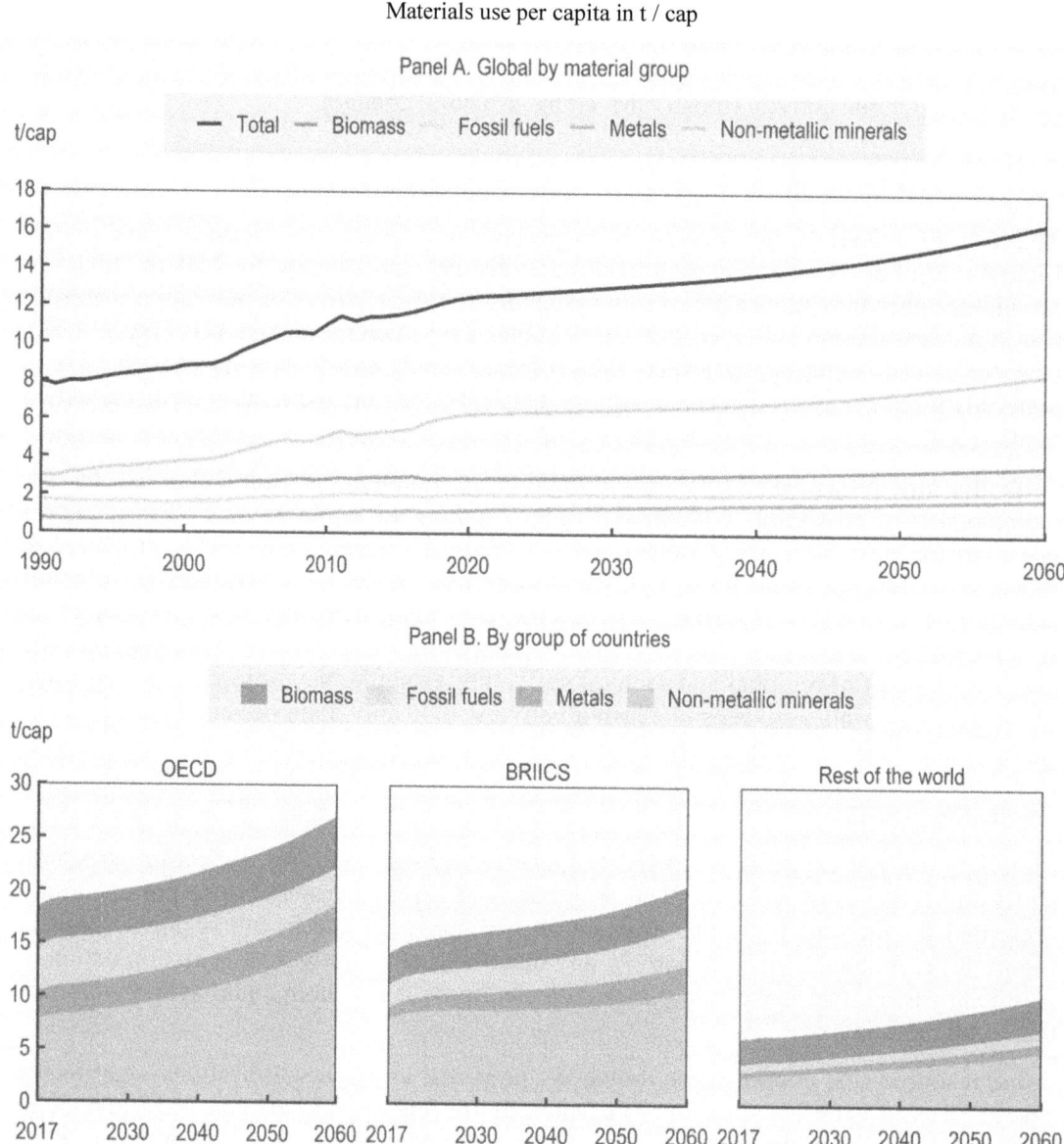

Figure 5.10. Global materials use per capita is projected to keep increasing over time

Source: OECD ENV-Linkages model.

StatLink https://doi.org/10.1787/888933885106

CHAPTER 5. PROJECTIONS OF PRIMARY MATERIALS USE | 133

Figure 5.11. Materials use per capita is projected to increase in most countries

Materials use per capita in t / cap

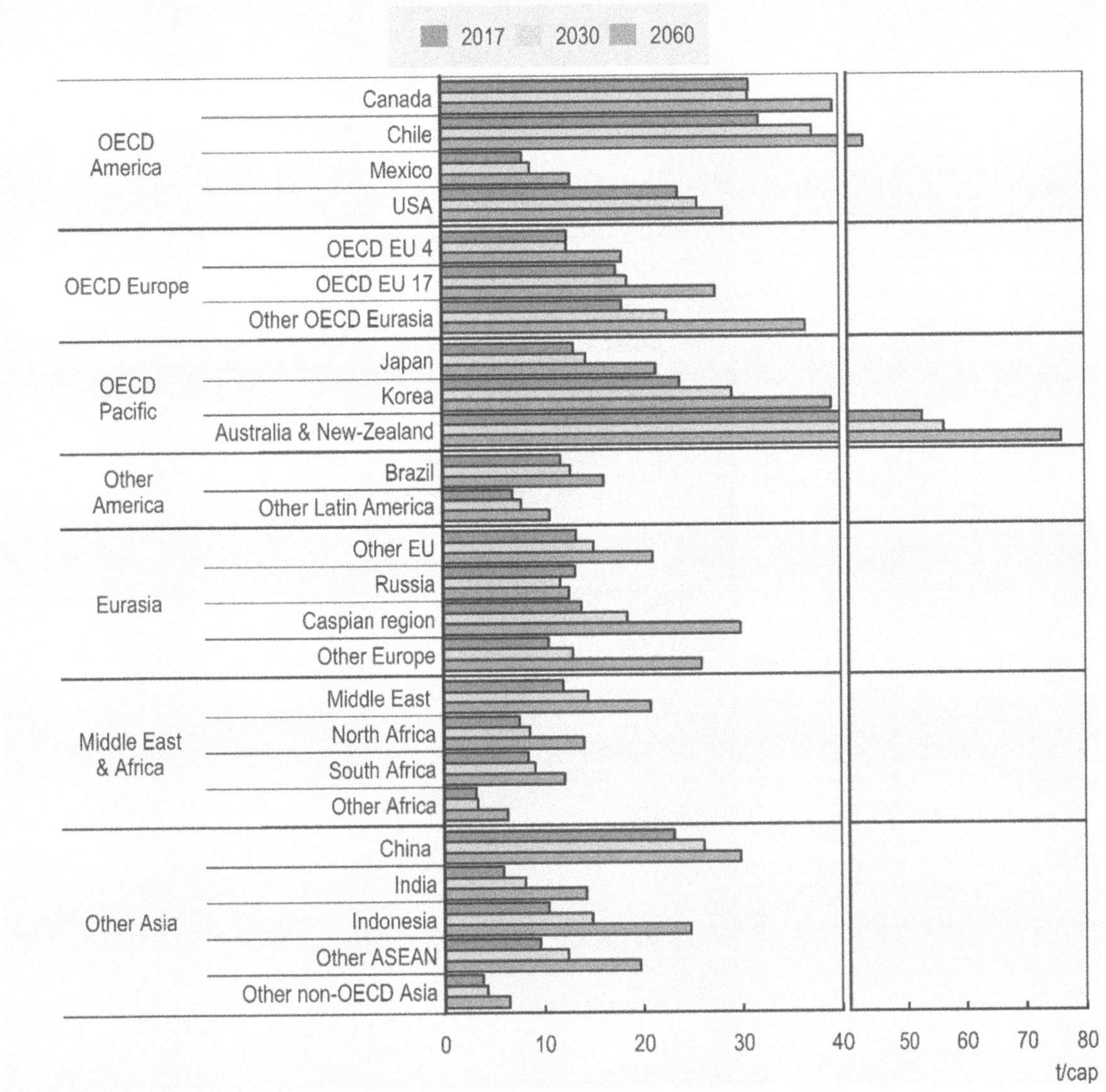

Note: In line with the methodology outlined in Chapter 2, materials that are domestically processed and then exported are attributed to the country of processing. This affects for instance the results for Australia and New Zealand.

Source: OECD ENV-Linkages model.

StatLink https://doi.org/10.1787/888933885125

Figure 5.12. Construction materials use per capita is projected to increase in most countries

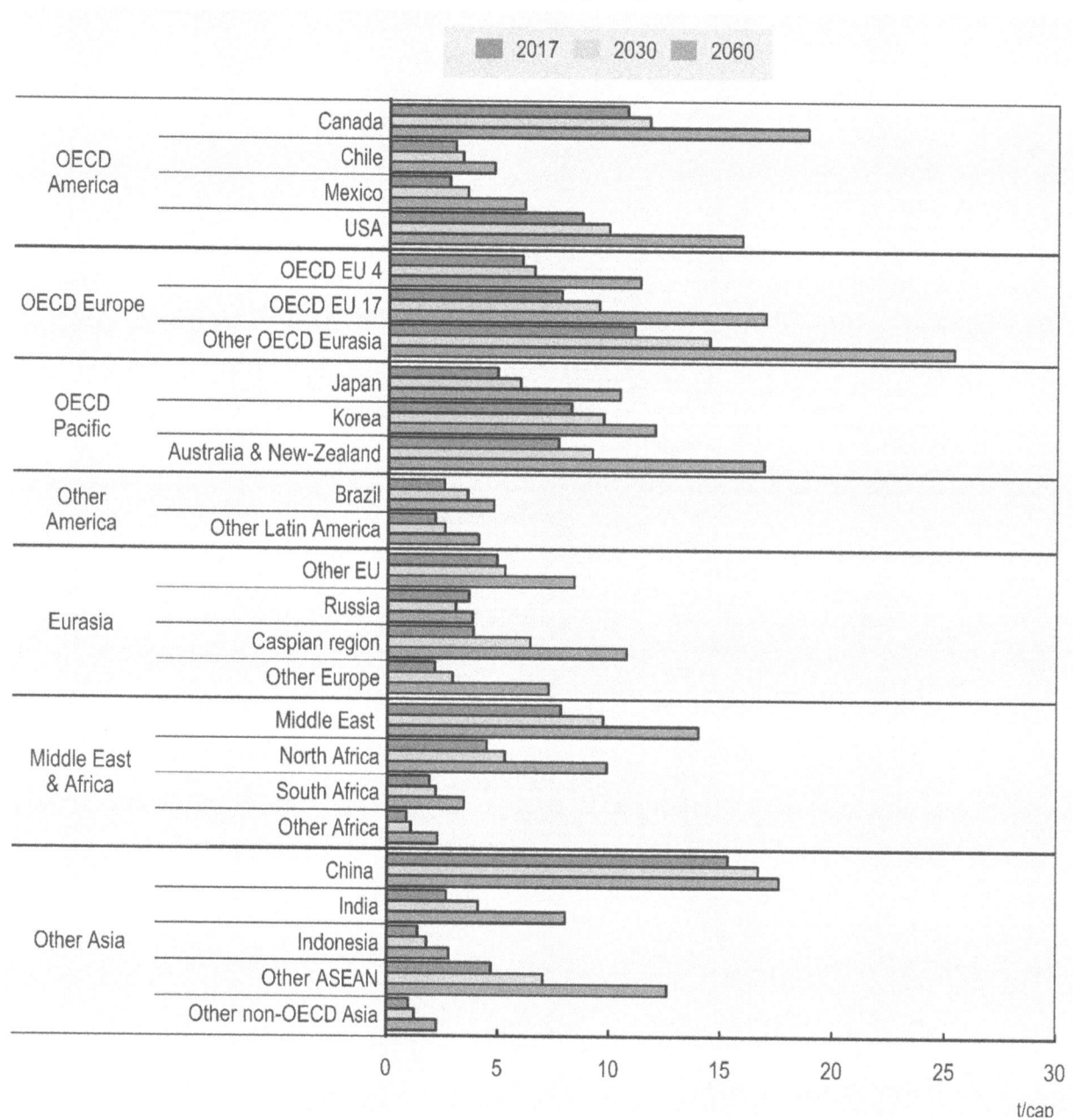

Source: OECD ENV-Linkages model.

StatLink https://doi.org/10.1787/888933885144

5.5. Socioeconomic scenarios are a source of uncertainty

Alternative assumptions about socioeconomic scenarios directly translate into changes in projections of materials use. Modelling the various alternative population and convergence rate scenarios presented in Section 3.4 in Chapter 3 reveals a range in projected materials use for 2060 between -20% and +20% from the central baseline scenario (Figure 5.13). This represents a variation of -33 to +32 Gt. Assumptions about economic convergence seem to be a more important source of uncertainty surrounding the use of materials than assumptions about population growth rates.

The regions with the largest projected materials use in 2060 in the central baseline scenario also have the largest uncertainty surrounding this projection. To some extent this reflects the way the alternative baselines are implemented: the same percentage variation on a larger starting value provides a larger variation in the projection. But economic dynamics also play a role here: a reduction (or increase) in growth rates in the first few years will affect savings and thus capital accumulation, leading the economy onto a permanently lower (or higher) growth path. From Figure 5.13 it also appears that the uncertainty range for non-metallic minerals is smaller for the OECD regions than the corresponding ranges for other materials. However, this is a visual effect caused by the very large range in materials use for the Other Asia region, which comprises many emerging economies.

All materials are equally affected by these alternative assumptions, except biomass whose changes are smaller (half the relative variation of other materials): in Figure 5.13 the scale of the x-axis is larger for biomass than for fossil fuels and metals, but the width of the uncertainty range is smaller. The low income elasticity for food plays a key role here: food demand largely scales with population, but much less with income.

The relatively narrow differences in ranges between materials are largely explained by the fact that all alternative scenario specifications stem from changing the assumptions on macroeconomic growth drivers; an uncertainty analysis focused more on structural change would logically have more differential effects on materials use.

The range of uncertainty on the projections of materials use, materials intensity and per-capita materials use are compared in Figure 5.14. Panel A shows the projected range for total materials use by region. The uncertainties scale with the size of the region, and are logically largest at the global level. The upper range of the projected total global materials use amounts to almost 200 Gt and is quite similar to UNEP's projection (2017[1]). Panel B shows materials intensity, i.e. materials use divided by GDP. The alternative baseline scenarios only cover macroeconomic uncertainties (population and income), but the underlying assumptions on materials productivity are identical to those of the central baseline scenario. Hence, the materials intensity projections are virtually identical across the range of alternative baselines and the uncertainty ranges are virtually invisible in the figure. At the global level, materials intensity in 2060 ranges between 0.42 and 0.43 USD/tonne, with the central baseline projection at 0.423 USD/tonne. This is of course not a realistic representation of the true uncertainty surrounding future materials intensity; for that a more elaborate uncertainty analysis would be required.

The uncertainty ranges for per-capita materials use are larger, as the variations in GDP projections directly affect this indicator. The largest uncertainties surround the projections for the BRIICS countries: the socioeconomic uncertainties are fairly large, and the central baseline scenario levels are higher than in the rest of the world region.

Figure 5.13. Uncertainties on population and income convergence strongly impact materials use projections

Uncertainty range on materials use in 2060 in Gt

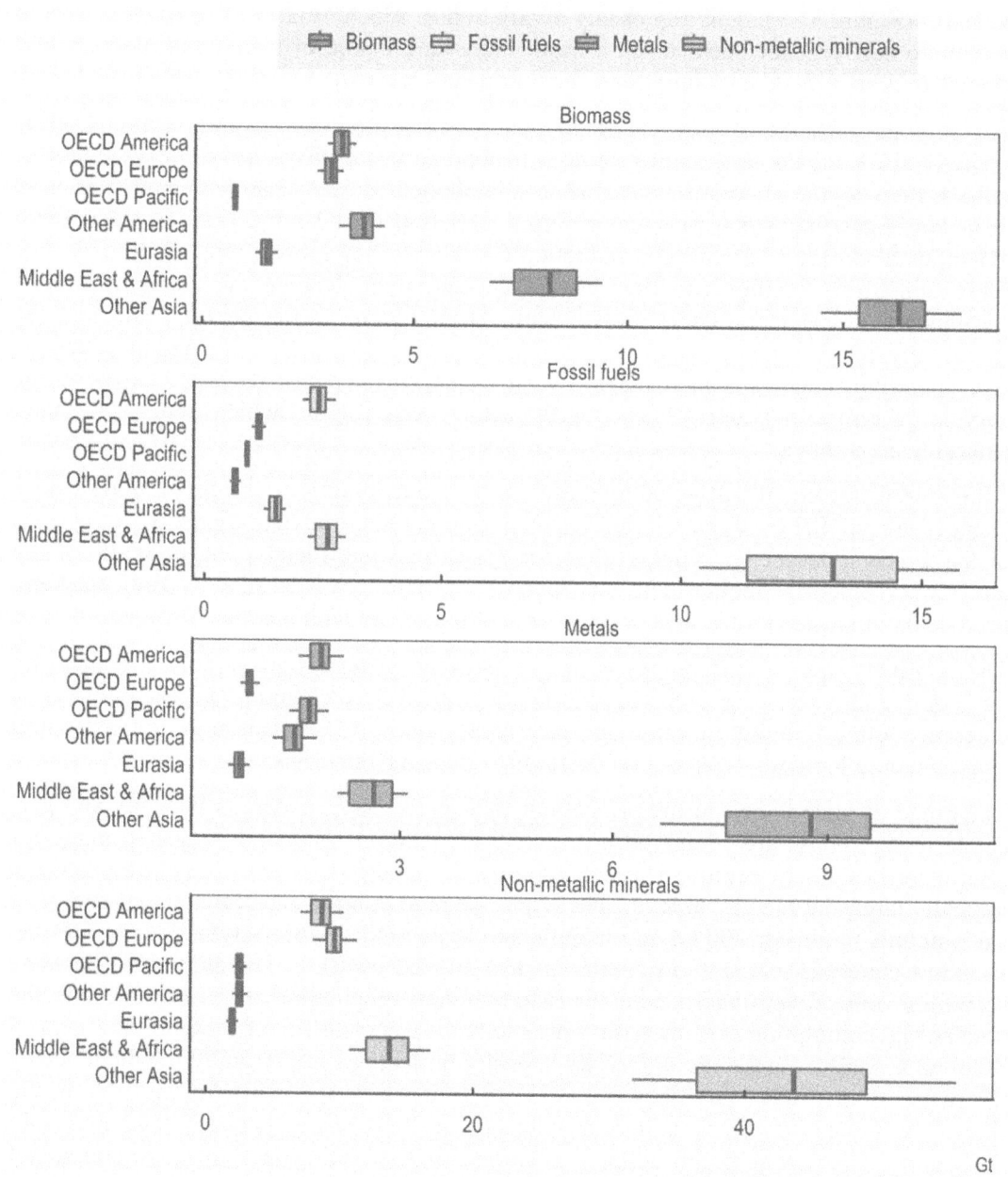

Note: The vertical black line shows the central baseline scenario value. The box shows the uncertainty range related to macroeconomic drivers (convergence assumptions), while the whiskers represent additional demographic uncertainty (population assumptions).
Source: OECD ENV-Linkages model.

StatLink https://doi.org/10.1787/888933885163

Figure 5.14. Materials intensity projections are less impacted by population and income convergence uncertainties

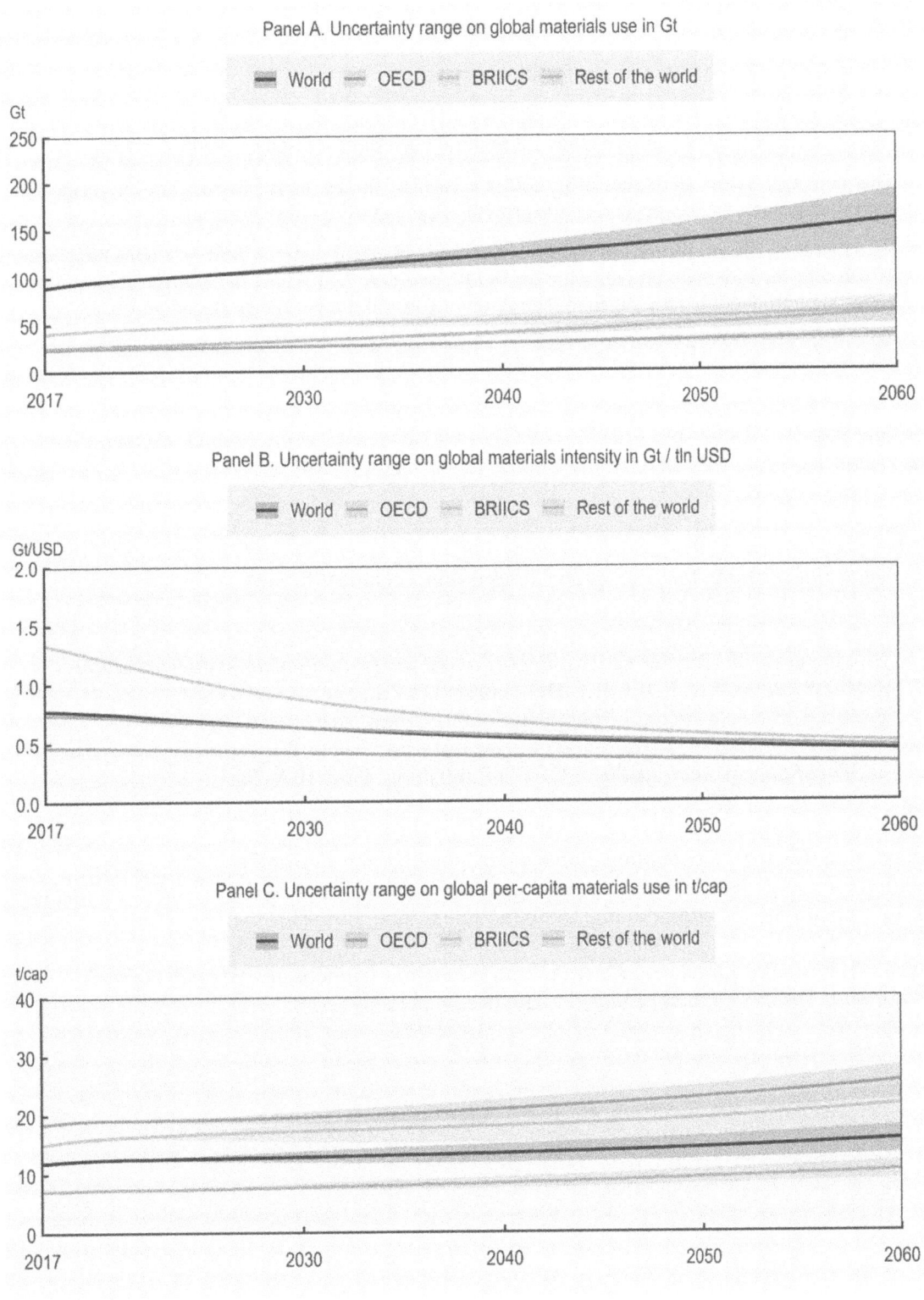

Source: OECD ENV-Linkages model.

StatLink https://doi.org/10.1787/888933885182

Notes

¹ Annex 5.A presents detailed results for all materials.

² Sand, gravel and crushed rocks constitute the bulk of concrete, while limestone enters in the composition of cement to aggregate them. Structural clays are for instance used for bricks.

³ This conclusion does not necessarily hold for each of the 60 materials that are aggregated in these broad groups, as they depend on the growth of sectors using these materials.

⁴ Note that these are projections of materials use, not embedded material flows for consumption as calculated in the raw material consumption (RMC, see Box 2.1). Thus, materials are allocated to the region of production, rather than to the region of consumption of the produced goods.

References

C. Böhringer and T.F. Rutherford (2015), *The Circular Economy - An Economic Impact Assessment*, Report to SIN-IZA, https://sunstiftungsfonds.files.wordpress.com/2015/06/report-circular-economy.pdf. [4]

Cambridge Econometrics (2014), *Study on modelling of the economic and environmental impacts of raw material consumption*, European Commission Technical Report, http://dx.doi.org/10.2779/74169. [3]

Hu, J., S. Moghayer and F. Reynès (2015), *Report about integrated scenario interpretation EXIOMOD/LPJmL results*, POLFREE Deliverable D3.7B., http://www.polfree.eu/publications/publications-2014/report-d37b. [6]

IEA (2017), *World Energy Outlook 2017*, OECD Publishing, Paris/IEA, Paris, http://dx.doi.org/10.1787/weo-2017-en. [2]

Schandl, H. et al. (2016), "Decoupling global environmental pressure and economic growth: scenarios for energy use, materials use and carbon emissions", *Journal of Cleaner Production*, Vol. 132, pp. 45-56, http://dx.doi.org/10.1016/j.jclepro.2015.06.100. [5]

UNEP (2017), *Resource Efficiency: Potential and Economic Implications. A report of the International Resource Panel.*, Ekins, P., Hughes, N., et al. [1]

Annex 5.A. Detailed results and supplementary materials for Chapter 5

Figure 5.A.1. Projections of materials use by sector

Materials use in Gt.

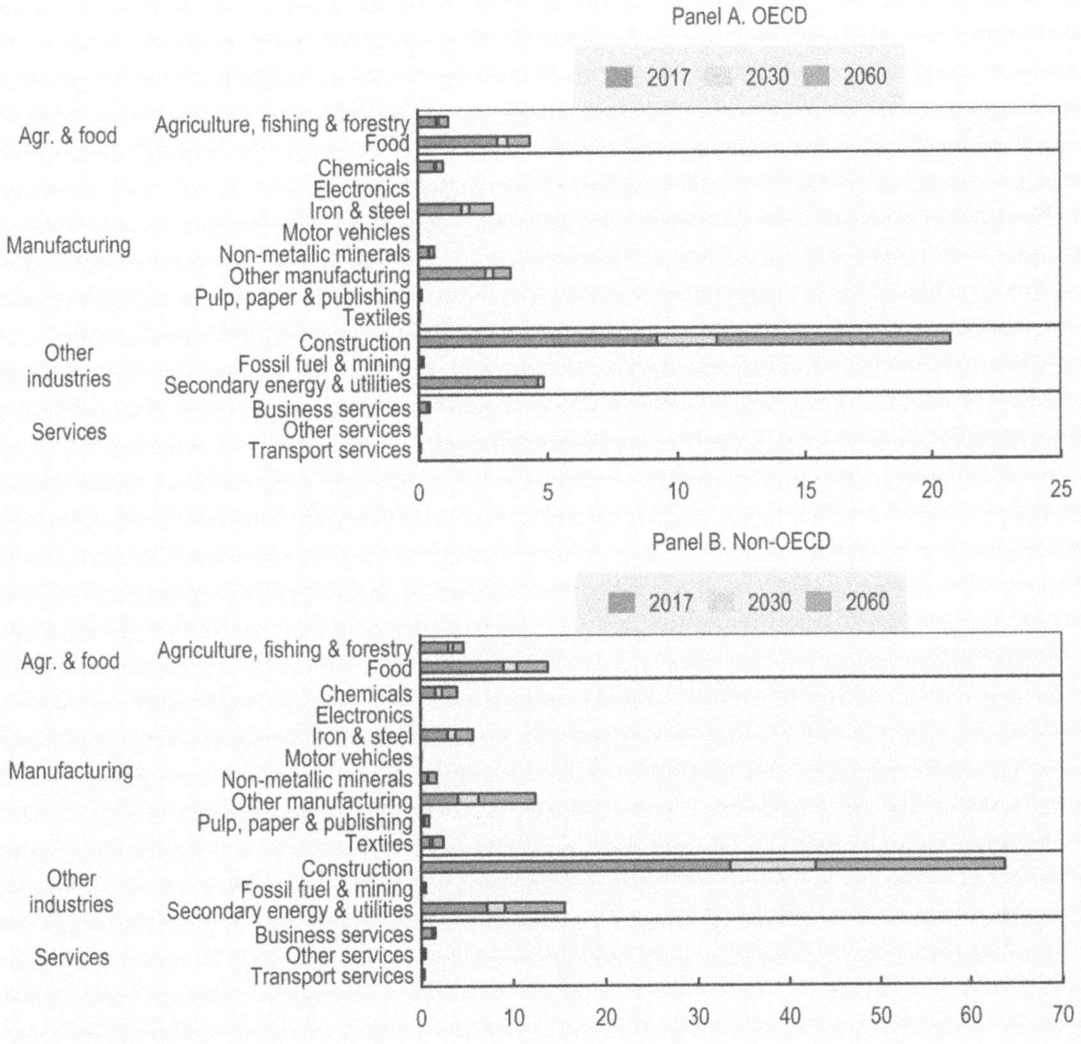

Note: The panels are not on the same scale.
Source: OECD ENV-Linkages model.

Chapter 6.

Projections of recycling and secondary materials

> *This chapter presents the projections of recycling and secondary materials use in the central baseline scenario. The first section presents an overview of the current situation on the use of secondary materials resulting from recycling. The second section presents projections of the recycling sector and the third section illustrates the impacts for secondary materials use to 2060, focusing on the competition between processing of primary and secondary materials.*

KEY MESSAGES

Recycling rates vary widely for different materials. Biomass and fossil fuels generally don't lend themselves to recycling, as they are expended or degraded when used. Many non-metallic minerals are too cheap or difficult to recycle. Some may however be downcycled, for building waste can be used for lower-value purposes such as road filler. For metals, recycling rates can be as high as 70%, such as for iron and steel. For several metals, competitive markets for recycled scrap exist. The recycled metal content in the economy is generally lower than the recycling rates, which is expected in growing economies: recycled content does not currently rise above 50%. As a result, the share of secondary materials in total materials use is limited, and is only significant for some metals.

Projections and trends

- Secondary materials (the result of processing recyclable waste into raw materials that can be used again) currently make up a modest part of total materials use. Many metals have substantial recycling rates, and scrap metals are used as secondary material. The share of secondary lead has surged in recent years to above 50%, while secondary steel has gradually declined to below 30%. Secondary shares for aluminium, zinc and copper are even lower. Recycling – and hence secondary materials use – is rare for non-metallic minerals; concrete is for example often used as low-value road filler.

- The recycling sector is projected to more than triple in size between 2017 and 2060 (see Figure 6.1). While both recycling and mining are projected to increase until 2030 at about the same pace, recycling is projected to grow more substantially from 2030 on. This is driven by the growth dynamics of developing countries: the high-growth phase triggers a boom in infrastructure, drawing largely on primary materials. As economies mature, however, and the increase of waste materials increases the availability of recyclable materials, the recycling sector begins to rise in importance.

- Recycling of materials is, however, projected to remain a small percentage of the total economy: the use of both primary and secondary materials is projected to rise in the central baseline scenario. Given the projected technology trends and unchanged policies, the supply of secondary materials is insufficient (or too expensive) to meet the demands of a growing economy.

- While the costs of recycling are projected to decline in comparison to mining new materials, the expansion of secondary materials is hampered by relatively high labour costs. As wages are projected to grow more rapidly than other production inputs, this implies a gradual decrease in the share of secondary materials in overall materials use, at least for non-ferrous metals.

Figure 6.1. The recycling sector is projected to outpace the mining sector

Growth of output between 2017 and 2060, index 1 in 2017

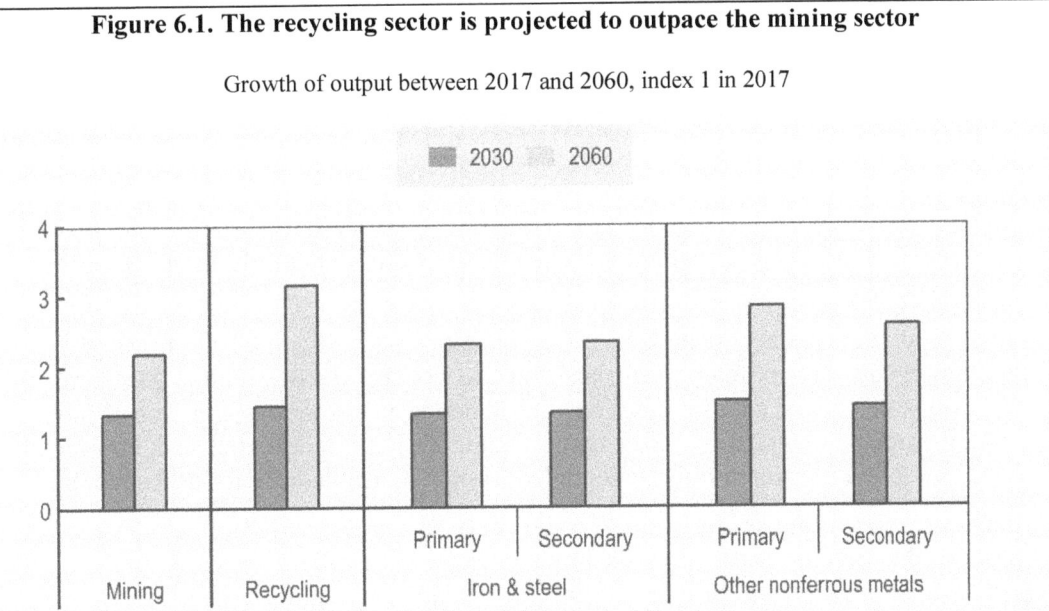

Source: OECD ENV-Linkages model.

StatLink https://doi.org/10.1787/888933885201

Areas of uncertainty

There are many uncertainties surrounding these projections. One modelling challenge is to develop consistent, comparable measures of raw primary materials (ores), raw secondary materials (scrap), and refined materials (from primary and secondary sources). The research quantified the likely impact of alternative assumptions about population growth and income convergence on the evolution of the output of the mining and recycling sectors, and on the share of secondary metals in total metals production. In both cases, this showed only a very minor impact as these are more driven by specific assumptions on structural change and technology developments than by macroeconomic conditions.

Policy implications

Further policy efforts are needed to broaden the scope of recycling to more materials, to increase the recycling rates of those materials where they are currently well below their potential, and to further increase the share of secondary materials use in total materials use. Chapter 5 has already highlighted that materials use needs to be further decoupled from economic activity, at both the sectoral and macroeconomic level. The technical and economic potential for increased recycling, together with the need to further reduce reliance on primary materials, implies that a policy mix is needed that can boost resource efficiency and stimulate the transition towards a more circular economy.

6.1. Secondary materials are only a modest part of total materials use

Secondary materials are the result of processing recyclable waste (scrap materials) into raw materials that can be used again. Secondary materials may be complete substitutes for primary materials, or may only be used in lower-value applications ("downcycling"): for instance when using recovered concrete as road filler. The processing and use of secondary materials are driven by the availability of scrap materials, and the cost-effectiveness of processing technologies using secondary materials input compared to primary materials.

In this report, secondary materials use projections are restricted to those that can substitute for the raw primary materials that are covered in the model. De facto, this implies only secondary metal projections are included in the analysis, as fossil fuels are expended through combustion, biomass is degraded after being used and non-metallic minerals are often recovered only in degraded form, e.g. usable only for downcycling.

Projections of recycling are for a much wider set of resources, and also includes recycling of processed materials. These processed materials are not explicitly covered in the current report, which focuses on primary versus secondary raw materials. While processed materials such as plastic and textiles can be recycled, unlike metals the raw material part can in general not be fully recovered without loss of value (crude oil cannot be recovered from plastic in an economically viable fashion, nor cotton from textile). They are captured in the model solely through the use of recycled products by the various processing sectors (for instance chemistry, furniture making and textiles).

One challenge in this modelling exercise resides in reconciling the physical measures of raw primary materials (ores), raw secondary materials (scrap), and refined materials (from primary and secondary sources). In some cases, the original mineral is recovered in the recycling process. In others, a processed form of the mineral is recycled. For example, the primary material iron ore is used to make steel, and steel scraps can be used as substitutes for iron ore in steel production. Sections 7.2 and 7.3 in Chapter 7 provide more detailed case studies for copper and iron and steel, respectively, and can provide further insights into recycling prospects for these specific metals.

While primary materials are usually measured in their raw form (as in Chapter 5: the weight of metal ores and non-metallic minerals, or the weight of biomass and fossil fuels), secondary materials are usually measured in their refined state. An alternative approach would be to present the volumes of scrap available and processed, but this would require a full stock accounting across the whole economy for all materials and all regions, which is not yet available. When comparing primary and secondary materials (for instance when assessing their environmental impacts in Chapter 8), the refined material measure is used.

Table 6.1 shows the recycling rates for a range of metals as detailed in UNEP (2011[1]). Recycling rates are presented for two metrics: *end-of-life (EoL) recycling rates*, i.e. the degree to which commodities are recycled at the end of life, and *recycled content*, i.e. the degree to which materials currently used in production consist of secondary materials. Recycling is at present mainly limited to materials that produce sufficient returns from recycling, which in many cases means that they must be present both in sufficient quantities and in sufficiently high concentrations in the supplied waste.

For many of the metal groups, recycling rates are substantial. The highest recycling rates are observed for chromium, followed by tin, iron and steel and platinum. Recycling rates for the other metals are generally much lower. High recycling rates do not necessarily translate into high shares of recycled content in existing commodities, not least due to long lifetimes of the products that contain metal. As production volumes keep increasing, inputs of both primary and secondary materials grow, and the recycled content remains substantially smaller than the EoL recycling rate.[1] A large potential for increasing recycling rates implies that the availability of recycled materials for secondary production could significantly increase without running into supply problems of secondary material.

Table 6.1. Estimates of current recycling rates and recycled content of metals

	End-of-life Recycling Rate (%)	Recycled Content (%)
Ferrous metals	70	40
Aluminium	55	35
Chromium	90	19
Copper	50	30
Gold	50	30
Manganese	53	37
Nickel	60	35
Silver	65	30
Tin	75	22
Zinc	40	23
Platinum group metal		
Iridium	25	17
Palladium	65	21
Platinum	70	20
Rhodium	55	40
Ruthenium	10	55
Other metals		
Antimony	20	5
Cobalt	32	68
Indium	0	38
Magnesium	39	33
Molybdenum	30	33
Niobium	53	22
Rhenium	17	60
Tantalum	5	20
Tungsten	46	40

Note: Recycled content refers to the secondary content of the refined metal production.
Source: Own calculations based on UNEP (2011[1]).

StatLink https://doi.org/10.1787/888933885866

Recycling is much rarer for non-metallic minerals. Concrete, which constitutes the bulk of non-metallic minerals in terms of weight, is often downcycled, e.g. concrete waste is used as road filler; while cement cannot be recycled. Some processed materials can also be recycled (glass is a good example) or downcycled (glass into glass wool as insulation materials).

Nonetheless, secondary materials make up a modest part of total materials use. Figure 6.2 presents the recent trends for selected metals. Significant increases have only been achieved for lead. For steel, the increases of the previous century were partially undone in the last decade; Section 7.3 in Chapter 7 delves into possible reasons for this. Shares of secondary processing have remained rather low for aluminium, copper, and especially zinc.

Figure 6.2. The share of secondary metals is very heterogeneous across selected metals

Shares of secondary metals in total global production (5-year moving averages)

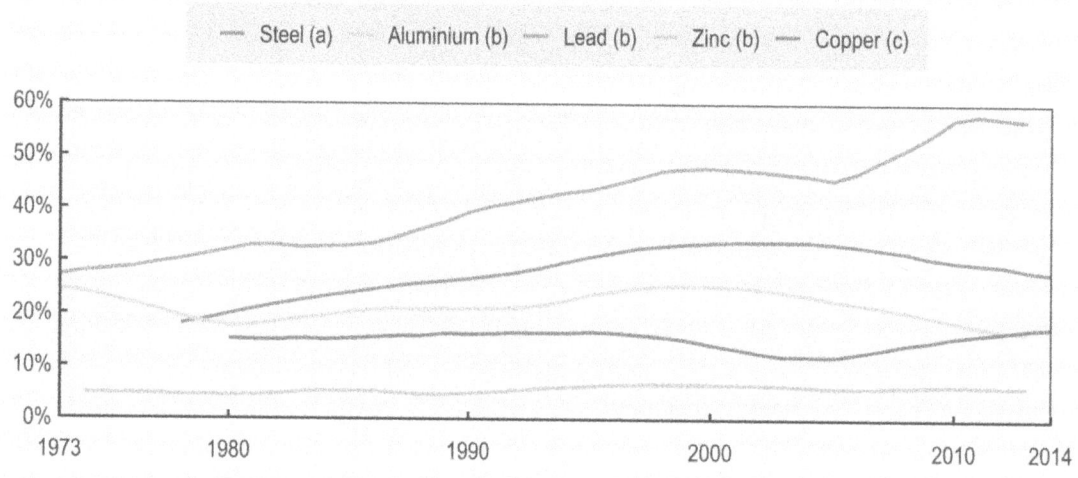

Note: The share of secondary metals is computed as the total production of refined metal from secondary sources in total refined material production.
Source: Own calculations based on (a) Worldsteel Association (2018[2]), (b) ABREE (2016[3]), and (c) USGS (2016[4]).

StatLink https://doi.org/10.1787/888933885220

Many different sectors use recycling as an input in their production (Figure 6.3). The recycling sector as modelled in this report corresponds to the recycling activity of all materials (not only metals, but also plastics, textiles, construction materials, glass, wood etc.). For instance, 10% of the output of the recycling sector goes to iron and steel primary production, 8% to the reprocessing of secondary steel. Also important are recycling activities for fruits and vegetables production (5%)[2], rubber and plastic (5%), and paper (5%). As a consequence, the recycling sector is a very wide activity which deals with varied streams of materials.

One challenge in correctly incorporating the recycling sector in the modelling framework is that currently a substantial share of recycling activities is provided by the informal sector, in particular in non-OECD countries. Box 6.1 explain the main issues and their consequences for the projections. This poses only limited drawbacks for the calculations, as the modelling framework directly projects secondary materials provision, without an explicit link to the source of the materials in the recycling process (as explained in Section 2.2 in Chapter 2).

Figure 6.3. Many sectors use recycling as input

Percentage of total demand for output of the recycling sector in 2017

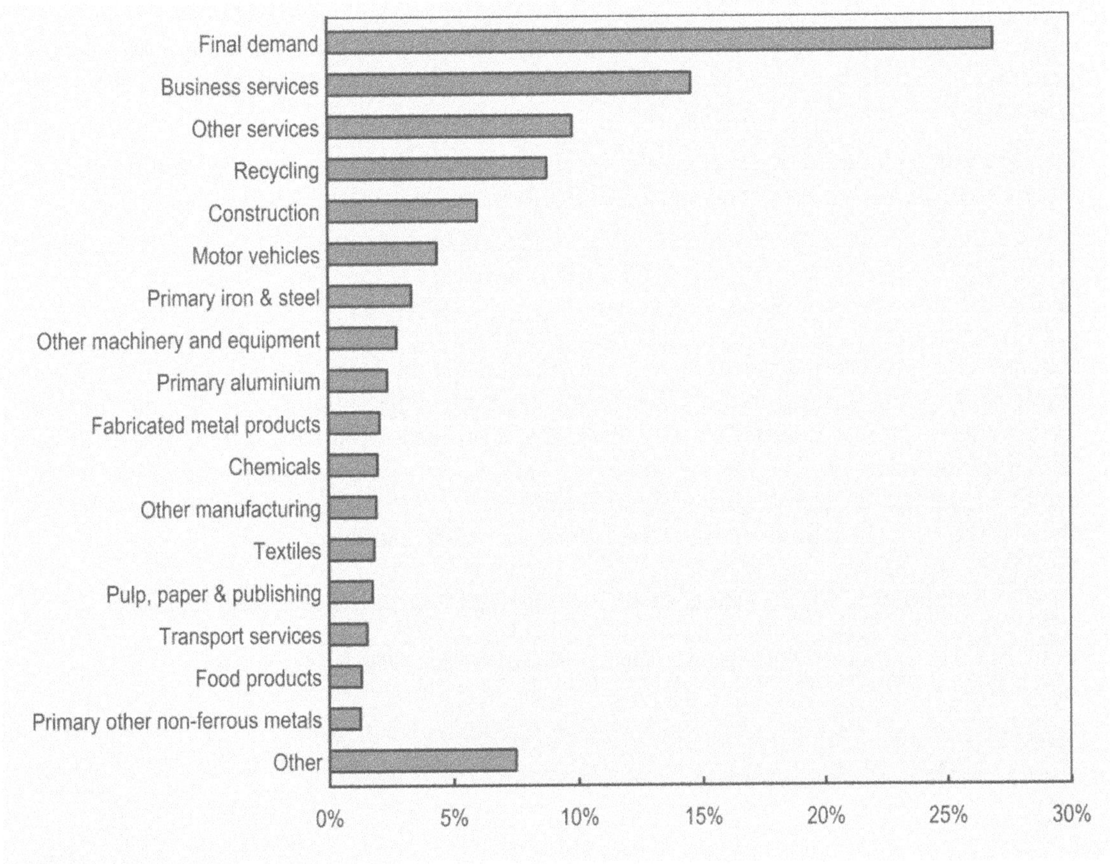

Sector	Percentage
Final demand	~27%
Business services	~14%
Other services	~9%
Recycling	~8%
Construction	~6%
Motor vehicles	~4%
Primary iron & steel	~3%
Other machinery and equipment	~3%
Primary aluminium	~2%
Fabricated metal products	~2%
Chemicals	~2%
Other manufacturing	~2%
Textiles	~2%
Pulp, paper & publishing	~2%
Transport services	~1.5%
Food products	~1%
Primary other non-ferrous metals	~1%
Other	~7%

Source: OECD ENV-Linkages model.

StatLink https://doi.org/10.1787/888933885239

Box 6.1. Representing informal sectors is challenging

Informal recycling activities are significant, as discussed in a recent UNEP report: "In many cities around the world there is a considerable presence of the informal sector in waste management, particularly in cities where there is no formal separate collection system for recyclable materials… The informal sector recovers, reuses or recycles valuable materials from waste and thereby contributes to sustainable resource management." (UNEP, 2016[5]).

The economic system of the OECD ENV-Linkages model used in this report is calibrated on the basis of countries' national account information at the base year. By definition, national accounts do not include informal economic activities (although they sometimes report an approximation of these). The modelling framework is thus unable to feature the activities of waste collection, material recycling and reprocessing that occur in the informal sector.

While this has little consequence for the main results presented in the report, this absence means that the production costs of the recycling sector are underestimated as some of the waste input expenses have been omitted. This will probably have little impact on the projection of the real costs of recycling in the long run.

The absence of the informal recycling activity also implies an underestimation of the recycling rates in the base year of the model. This could affect the prospects for increasing recycling rates in the future.

Nevertheless, as countries develop, the informal sector share can be expected to decline progressively, causing these underestimations to vanish.

6.2. Recycling is projected to triple

The recycling sector is projected to more than triple in size between 2017 and 2060 (Figure 6.4). In contrast, mining activities approximately double during that period, growing more slowly than GDP. This indicates a projected increase in the weight of the recycling sector in the economy. However, both sectors remain small compared to the size of the global economy: mining decreases from 0.7% to 0.6% of total output while the share of recycling maintains more or less constant share just below 0.04%.[3]

Figure 6.4. The recycling sector is projected to outpace the mining sector

Growth of global sectoral output between 2017 and 2060, index 1 in 2017

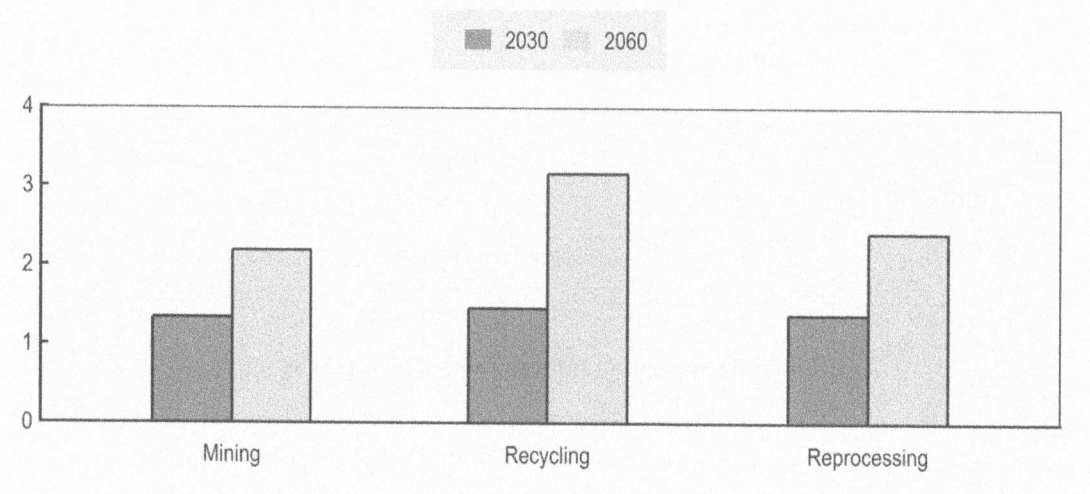

Source: OECD ENV-Linkages model.

StatLink https://doi.org/10.1787/888933885258

The regional dynamics largely follow the global evolution. Figure 6.5 illustrates that the recycling sector is projected to grow faster than the mining sector in all represented regions. Furthermore, in most regions the projected growth of the reprocessing sectors is faster than the growth of mining as well; regions with strong projected growth in recycling are also projected to have rapidly growing reprocessing sectors.

CHAPTER 6. PROJECTIONS OF RECYCLING AND SECONDARY MATERIALS | 149

Figure 6.5. In almost all regions, recycling is projected to grow more rapidly than mining

Growth of output between 2017 and 2060, index 1 in 2017

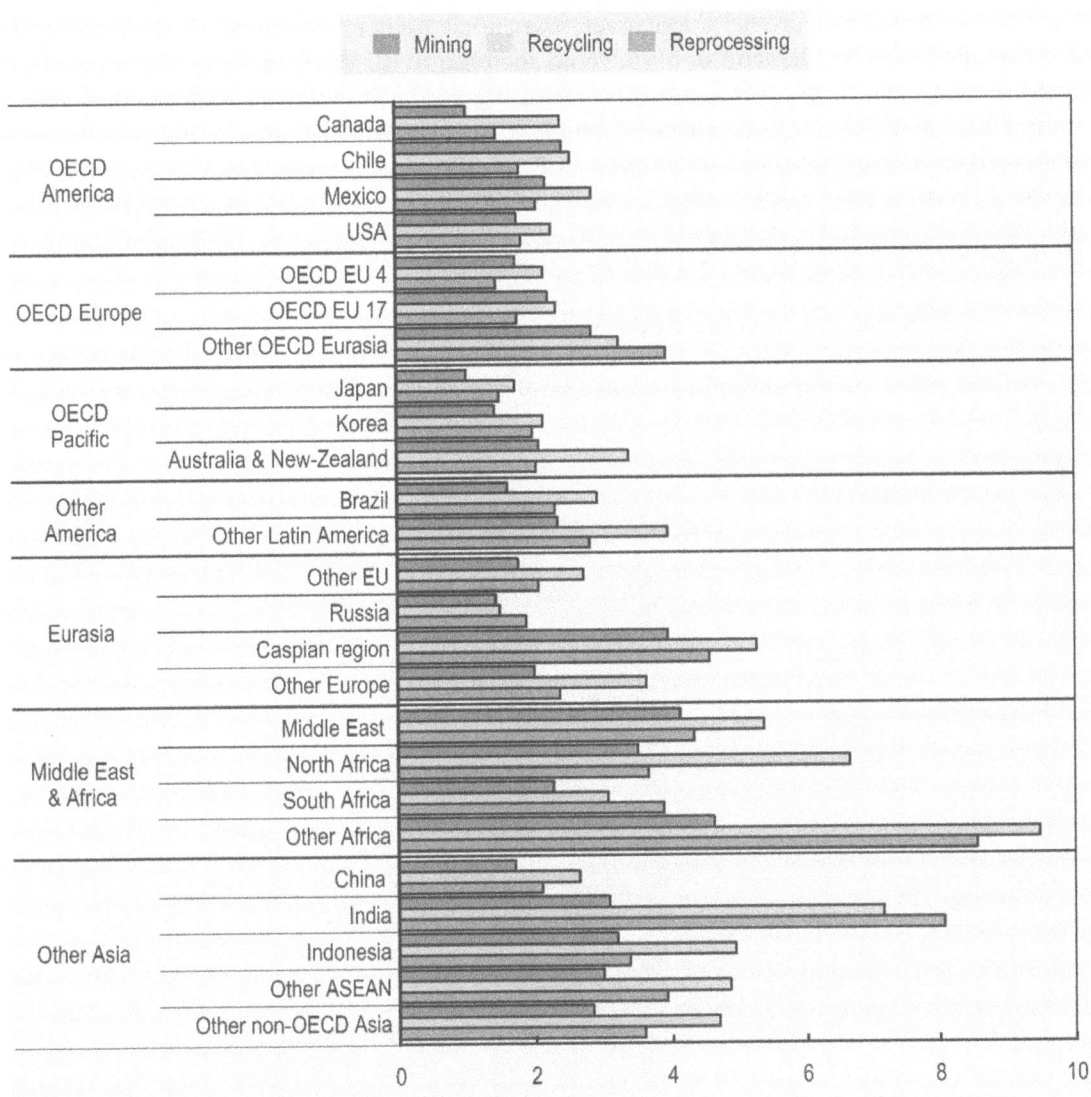

Note: See Box 2.1 in Chapter 2 for definitions of recycling and reprocessing and Table 2.1 in Chapter 2 for regional definitions. In particular, OECD EU 4 includes France, Germany, Italy and the United Kingdom. OECD EU 17 includes the other 17 OECD EU member states. Other OECD Eurasia includes the EFTA countries as well as Israel and Turkey. Other EU includes EU member states that are not OECD members. Other Europe includes non-OECD, non-EU European countries excluding Russia. Other Africa includes all of Sub-Saharan Africa excluding South Africa. Other non-OECD Asia includes non-OECD Asian countries excluding China, India, ASEAN and Caspian countries.
Source: OECD ENV-Linkages model.

StatLink https://doi.org/10.1787/888933885277

However, the speed of growth of mining, recycling and reprocessing is variable across countries. While OECD countries are projected to roughly double the size of the recycling sector, the mining sector more than doubles in specific resource-exporting countries (Chile, Mexico, Other OECD Eurasia, Australia) and grows more modestly in other OECD countries. In most OECD countries, but not Other OECD Eurasia (which includes the EFTA countries, plus Israel and Turkey), the reprocessing sector growth roughly follows that of mining.

Similarly, the recycling sector in non-OECD countries is projected to increase faster than the mining sector (Figure 6.5). However, the much higher growth rates of economic activity and materials demand imply a rapid increase in all three sectors. The recycling sector is projected to grow fastest in the emerging and developing economies (not least India and the African regions except South Africa). The metal reprocessing sectors are also projected to grow faster than mining in most non-OECD countries, and in India are projected to outpace even the rapidly increasing recycling sector. At the global level, the mining, recycling and reprocessing sectors are projected to multiply their output levels between 2017 and 2060 by 2.2, 3.2 and 2.4, respectively.

Figure 6.6. The recycling sector is projected to remain small in all regions

Share of the sectors in the economy in 2060, sorted by the highest share of the recycling sector

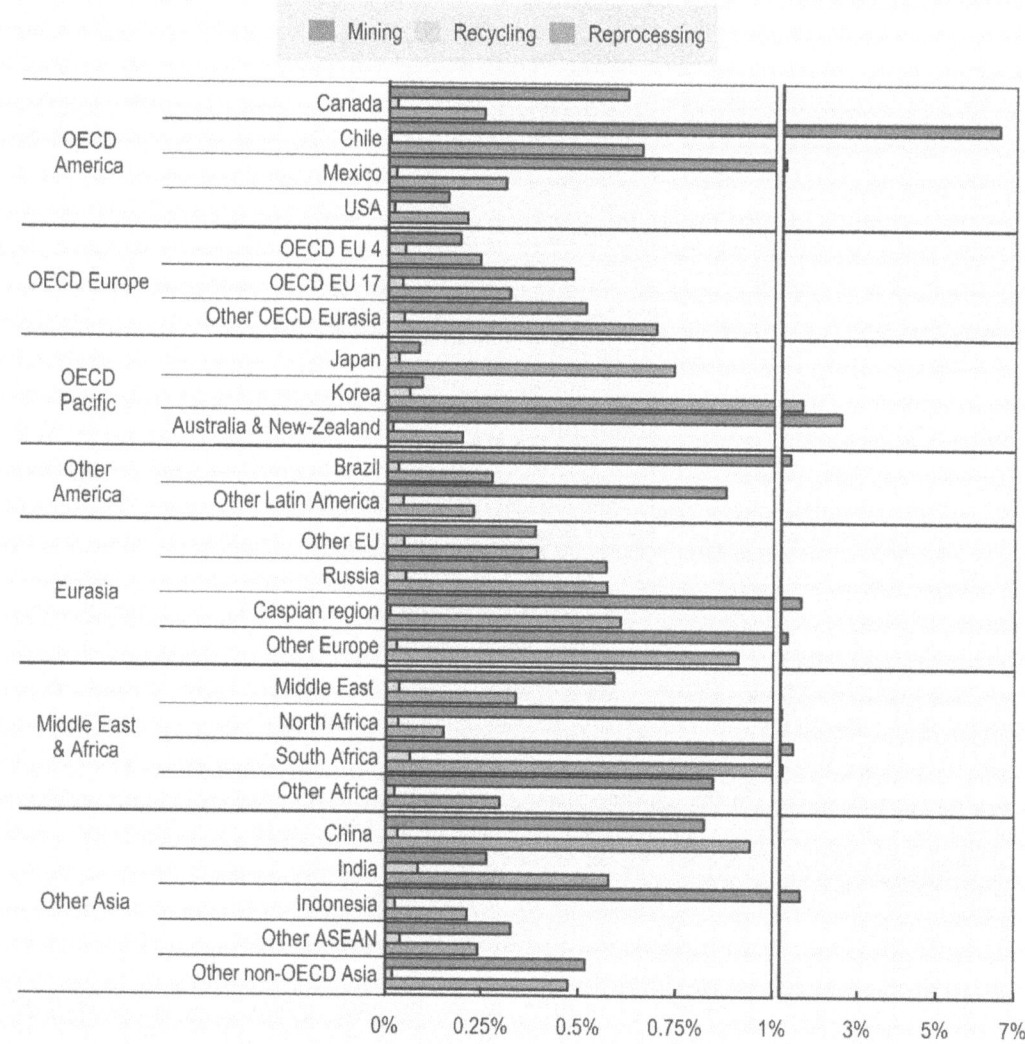

Source: OECD ENV-Linkages model.

StatLink https://doi.org/10.1787/888933885296

Nonetheless, the projected share of the recycling sector in 2060 still remains very small in comparison to the overall size of the economy (Figure 6.6). The share of the recycling sector in the total value of the economy is projected to remain below 0.1% in all countries, with India coming closest to that level.

The share of the metal reprocessing sector is, however, projected to be similar to the share of the mining industry globally (0.5%). However, this share varies widely by country (Figure 6.6). Resource-rich economies rely on mining exports and their mining sector thus represents a higher share of value added (see for instance Chile or Australia). These geographical differences entail opportunities for resource importing countries to use reprocessing as a source of growth and employment.

Box 6.2 discusses the plausibility of increased recycling in a baseline scenario from a technical perspective. For most materials, current end-of-life recycling rates are below their technical and economic potential, and there is thus room to increase recycling rates, even in absence of policies to stimulate recycling.

Box 6.2. The potential for increasing recycling rates

In the industrially developed world the metallurgical step of recycling for the most important metals can be considered optimized given current production costs, market incentives and policies. As reported in UNEP (2013$_{[6]}$): *"Ingenuity in metallurgy has helped the industry to drive the efficiency of recycling of ferrous and base metals (e.g. steel, stainless steel, aluminium, copper, zinc, lead, nickel, tin) ever closer to the limits that are permitted by physics and thermodynamics"*.

More non-ferrous metals (i.e. aluminium and copper) could be extracted from bottom ash. UNEP (2011$_{[11]}$) estimates that the current extraction from waste streams of 130 kt (of which 65% aluminium) could be tripled in 2020 as a result of better process technology and an increase in waste-to-energy plants.

The extent of the collection is far from optimal in most industrialized countries : *"In many cases (sometimes despite legislation) small articles or Waste Electrical and Electronic Equipment are not collected separately for recycling but disposed of with Municipal Solid Waste"* (UNEP, 2013$_{[6]}$). For smaller electronic equipment, a doubling of the collection rate could be achieved. Larger equipment (e.g. cars or industrial production equipment) already has a considerably higher collection rate, so there is no such potential for improvement. Given the different shares of metal use in smaller or larger electronic equipment, the potential for increased recycling from electronics is particularly high for cobalt, gallium and indium; average for gold, silver, palladium and platinum; and moderate to poor for rare earth elements.

In view of these arguments, the recycling percentages for aluminium, cobalt, indium and gallium can be assumed to be below their technical potential. This also applies to more commonly used metals like zinc, tin and lead.

Secondary materials provided by the recycling sector depend on the availability of waste. Waste streams are not modelling in this report, so the link between waste generation and recycling cannot be made (see Section 2.2 in Chapter 2). The World Bank has, however, used the central baseline scenario presented in Chapter 3 to project future municipal waste. These waste projections are briefly presented in Box 6.3.

Box 6.3. Municipal solid waste is a growing issue

Waste is a growing global issue with serious consequences for the environment and public health, when not managed properly. Waste management is of particular concern in urban areas, where the high population density leads to both high level of waste generation and strong potential impacts on health due to the proximity of inhabitants. The modelling framework used in this report does not allow quantification of future waste generation. However, the World Bank (2018[7]) makes projections of future Municipal Solid Waste (MSW) – defined as residential, commercial and institutional waste – that are based on the central baseline scenario presented in Chapter 3.

According to the World Bank projections (2018[7]), 2.1 Gt of MSW is currently generated worldwide; this averages 0.77 kg per person per day at the global level, but has a wide range in different countries – from 0.12 kg to 4.39 kg. Although high-income countries only account for 16% of the world population, they generate approximately 32% of the world MSW (680 Mt). A conservative estimation suggests that at least 25% of MSW is not managed in an environmentally safe manner.

Global MSW is projected to grow to 3.8 Gt by 2050 (see figure below). MSW generation per capita in high-income countries is projected to increase only slightly (by 11%) by 2050, while in low-income countries, it is expected to nearly triple. MSW is shown to increase at a faster rate for low income countries but to slow down as their income increases. The total MSW generation quantity in low-income countries is expected to increase by more than a multiple of six by 2050.

Figure 6.7. Projected municipal waste generation increases in all regions to 2050

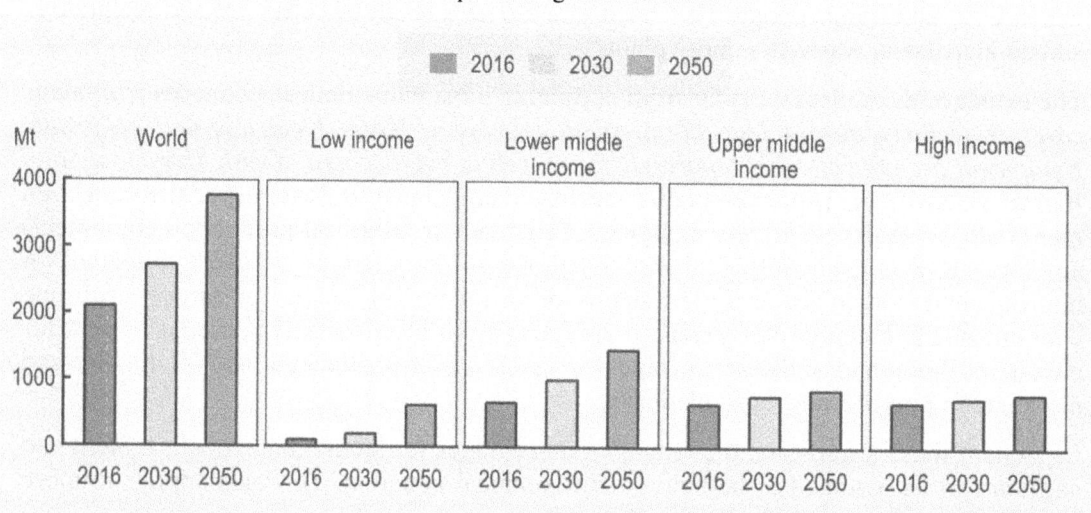

Source: World Bank (2018[7]).

6.3. Secondary metal production is projected to grow as fast as primary metal production

In this report, the projections of secondary metals use are driven by demand. The different metals processing sectors depicted in the model can produce the same good using two types of material inputs: (i) primary materials, from the processing of mined, extracted metals, and (ii) secondary metals, from the recycling of waste scrap. The process that uses primary metals is usually more energy and capital intensive and less labour intensive than the process using secondary metals (cf. Figure 4.11 in Chapter 4). The secondary metal production is projected to increase at roughly the same pace to 2060 than primary, as seen in Figure 6.8.

Figure 6.8. The share of secondary metal production is projected to remain roughly unchanged until 2060

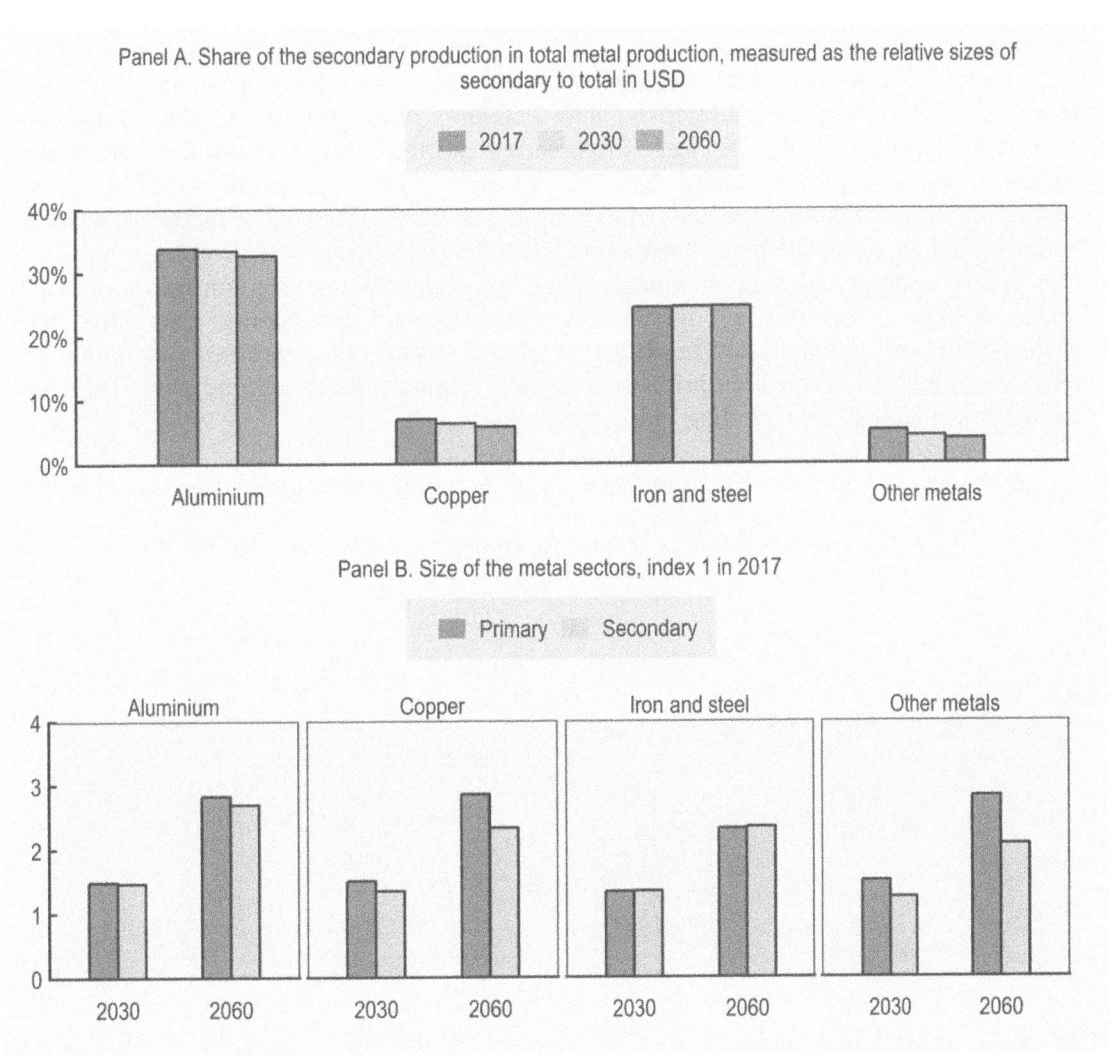

Source: OECD ENV-Linkages model.

StatLink https://doi.org/10.1787/888933885334

For copper and other non-ferrous metals the share secondary is projected to slightly decline. For iron and steel, where the secondary production is relatively more mature, and cost differentials between primary and secondary are smaller, the projections show no significant difference between both.

As discussed above, the needs of emerging and developing economies for materials is so large that they need the primary materials to build stocks. That will in turn increase the availability of scrap stocks for later recycling (see Chapter 7 for more insights on the future availability of scrap metals for copper and iron and steel), thus opening up the possibility to shift towards more production based on secondary materials in response to policies.

As described in Chapter 2, relative price differentials drive the dynamics of competition between primary and secondary materials. Figure 6.9 shows how these price changes affect the relative production costs of secondary metals versus primary metals production. On the one hand the prices of metal ores and scraps change over time in favour of secondary metal production (cf. Figure 4.10 in Chapter 4). But this effect is dominated by the relative increase in wages compared to capital costs. As primary production of non-ferrous metals is more capital intensive and secondary production more labour intensive (Figure 4.11 in Chapter 4), this wage increase reduces the growth potential of secondary metals production in the central baseline scenario. Thus, the evolution of the cost competitiveness of secondary metal production compared to primary metal production favours primary sources for non-ferrous metals throughout the whole projection horizon. However, the projected medium run dynamics of copper and other non-ferrous metals imply growth in primary production, as emerging and developing economies are projected to grow strongly and build their material stocks. The increased maturation of these economies then relies on primary materials, while at the same time paves the way for scrap availability for recycling in the long run.

Figure 6.9. The relative price of secondary non-ferrous metals is projected to increase

Evolution of the ratio of secondary metal price to primary metal price compared to 2017

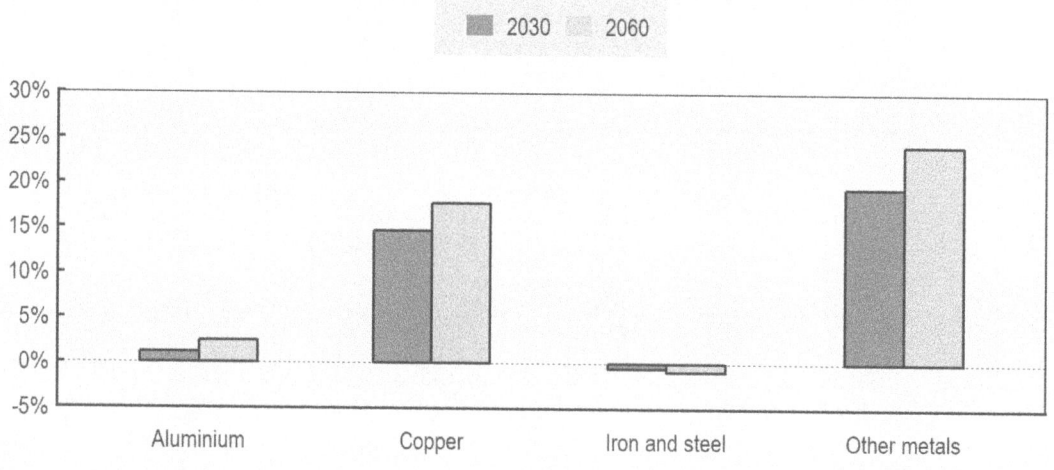

Source: OECD ENV-Linkages model.

StatLink https://doi.org/10.1787/888933885353

6.4. Uncertainty surrounds the recycling and secondary materials projections

There are many uncertainties surrounding these projections. For example, the assumed substitutability between primary and secondary materials is based on relatively weak empirical evidence. Similarly, the evolution of the costs of recycling vis-à-vis mining hinges on assumptions regarding technological change and price developments. A full analysis of these uncertainties is beyond reach for this report. But the impact of the alternative assumptions about population growth and income convergence as described in Section 3.4 in Chapter 3 can be quantified. These quantitative results should not be seen as indicative of the full uncertainty range surrounding the central baseline projections of recycling and secondary materials use, but only serve to highlight the role of these socioeconomic drivers.

Figure 6.10 highlights how the socioeconomic uncertainties could affect the evolution of the output of the mining and recycling sectors. The graph shows that the output of these sectors is about as sensitive as GDP. This implies that the effect of these uncertainties on structure of the economy is very limited.

Figure 6.10. Mining and recycling output vary with population and income convergence assumptions in proportion to GDP

Sectoral output growth; index 1 in 2017

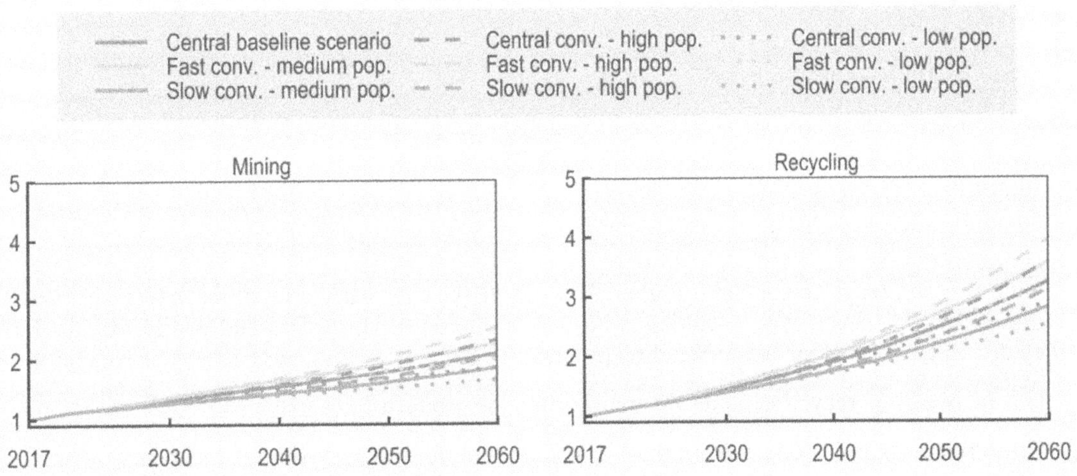

Source: OECD ENV-Linkages model.

StatLink https://doi.org/10.1787/888933885372

The alternative assumptions on population and income convergence also have only a very minor impact on the projected evolution of the share of secondary metals in total metals production, as shown in Figure 6.11. These socioeconomic assumptions are not strong enough to change the overarching trend that the share of secondary gradually declines for aluminium, copper and other metals, while it stays more or less stable for iron and steel.

Of course, changes in modelling assumptions that directly affect the trade-off between primary and secondary materials, or the evolution of recycling and mining sectors, are likely to have a much larger impact. These include the elasticity of substitution between primary and secondary production processes and assumptions that affect relative price changes, i.e. those on structural change and technology development in the mining, recycling and reprocessing sectors. Changes in policies will also have a significant impact on these trends.

Figure 6.11. The share of secondary metals is projected to change little under alternative population and income convergence assumptions.

Evolution of the share of secondary metal in percentage of total production

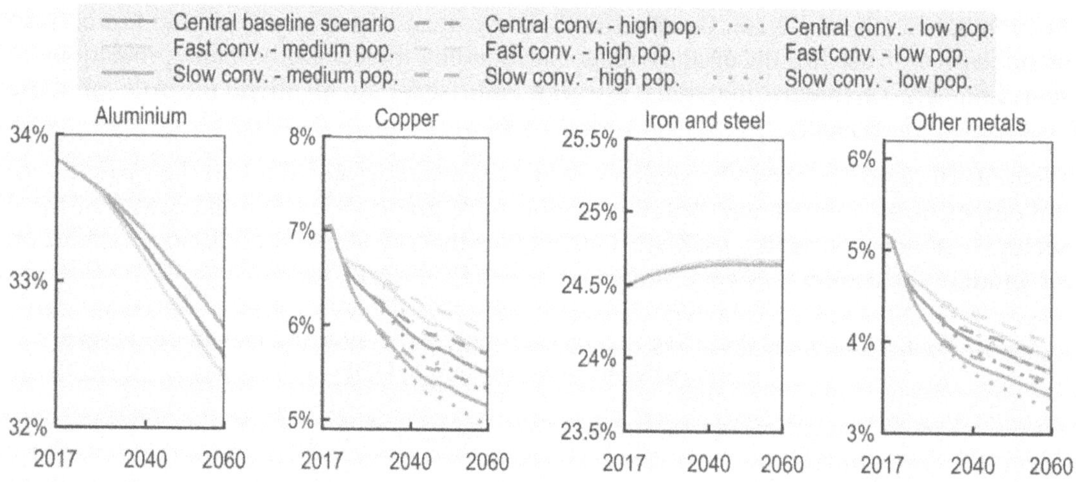

Source: OECD ENV-Linkages model.

StatLink https://doi.org/10.1787/888933885391

Notes

[1] In some cases, recycled content is higher. This can be the case when a lot of material is recovered at the production stage: these materials will then not be counted towards end-of-life recycling as they do not go through the waste handling process.

[2] This includes e.g. recycling boxes for transporting the produce.

[3] These figures only include extraction of primary materials and provision of secondary materials, not their (re-)processing which carried out by the corresponding industrial sectors.

References

ABREE (2016), *Resources and Energy Statistics*, https://industry.gov.au/Office-of-the-Chief-Economist/Publications/Pages/Resources-and-energy-statistics.aspx (accessed on 18 May 2018). [3]

Stadler, K. et al. (2018), "EXIOBASE 3: Developing a Time Series of Detailed Environmentally Extended Multi-Regional Input-Output Tables", *Journal of Industrial Ecology*, http://dx.doi.org/10.1111/jiec.12715. [8]

UNEP (2016), *Global Waste Management Outlook*, UN, New York, http://dx.doi.org/10.18356/765baec0-en. [5]

UNEP (2013), *Metal Recycling – Opportunitites, Limits, Infrastructure (Full report)*, United Nations Environment Programme, http://www.resourcepanel.org/sites/default/files/documents/document/media/e-book_metals_report2b_recyclingopportunities_130919.pdf (accessed on 18 May 2018). [6]

UNEP (2011), *Recycling Rates of Metals: a status report*, International Resource Panel, http://wedocs.unep.org/bitstream/handle/20.500.11822/8702/-Recycling%20rates%20of%20metals%3a%20A%20status%20report-2011Recycling_Rates.pdf?sequence=3&isAllowed=y (accessed on 16 May 2018). [1]

USGS (2016), *USGS Minerals Information: Copper*, https://minerals.usgs.gov/minerals/pubs/commodity/copper/index.html#myb (accessed on 18 May 2018). [4]

World Bank (2018), *What a Waste 2.0: A Global Snapshot of Solid Waste Management to 2050*, World Bank, Washington, DC. [7]

Worldsteel (2018), *Worldsteel*, https://www.worldsteel.org/ (accessed on 18 May 2018). [2]

Annex 6.A. Detailed results and supplementary materials

The recycling sector was split from the manufacturing sector in the GTAP database using the structure from the Exiobase database (see Figure 6.A.1). Panel A shows the production structure of the recycling service: most of the cost structure of the sector consists of labour costs (27 %), and capital costs (21 %). Furthermore, this sector overall benefits from no taxes applied to it. Furthermore, 18 % of production costs are dedicated to Services while 8 % are dedicated to self-consumption in the recycling sector itself. Finally, land transport is a key input as goods need to be collected.

Interestingly, the tax rates are very heterogeneous across countries for the recycling sector (see Figure 2.A.1). While input tax rates are positive for all countries – ranging from -0.1 % to 10.2 % – other production tax rates are very small. A remarkable figure is China with a very high subsidy, leading to a -13.6 % tax rate. Other countries range from -0.9 % to 1.5 %. The high subsidy in China explains the global tax rate close to zero (in Figure 6.A.1).

Figure 6.A.1, Panel B depicts the demand structure. While 10 % goes to Investment, and 4 % to households, the largest consumer is the Iron and steel processing (primary for 10 % as well as secondary for 8 %). Self-consumption also constitutes the end point of 8 % of the sector production. Aside from steel, the 'Vegetable and fruits' sector is one of the big consumer of this service with 5 % (for the pallet boxes?). The governmental services uses about 4 %, while plastics, paper and textiles appear next (respectively 4 %, 4 %, and 3 %). Construction only uses about 2 %, while at the same levels as 'Fabricated metal products', 'Aluminium Primary' and 'Aluminium Secondary'.

Figure 6.A.1. Global production and demand structure of the recycling sector

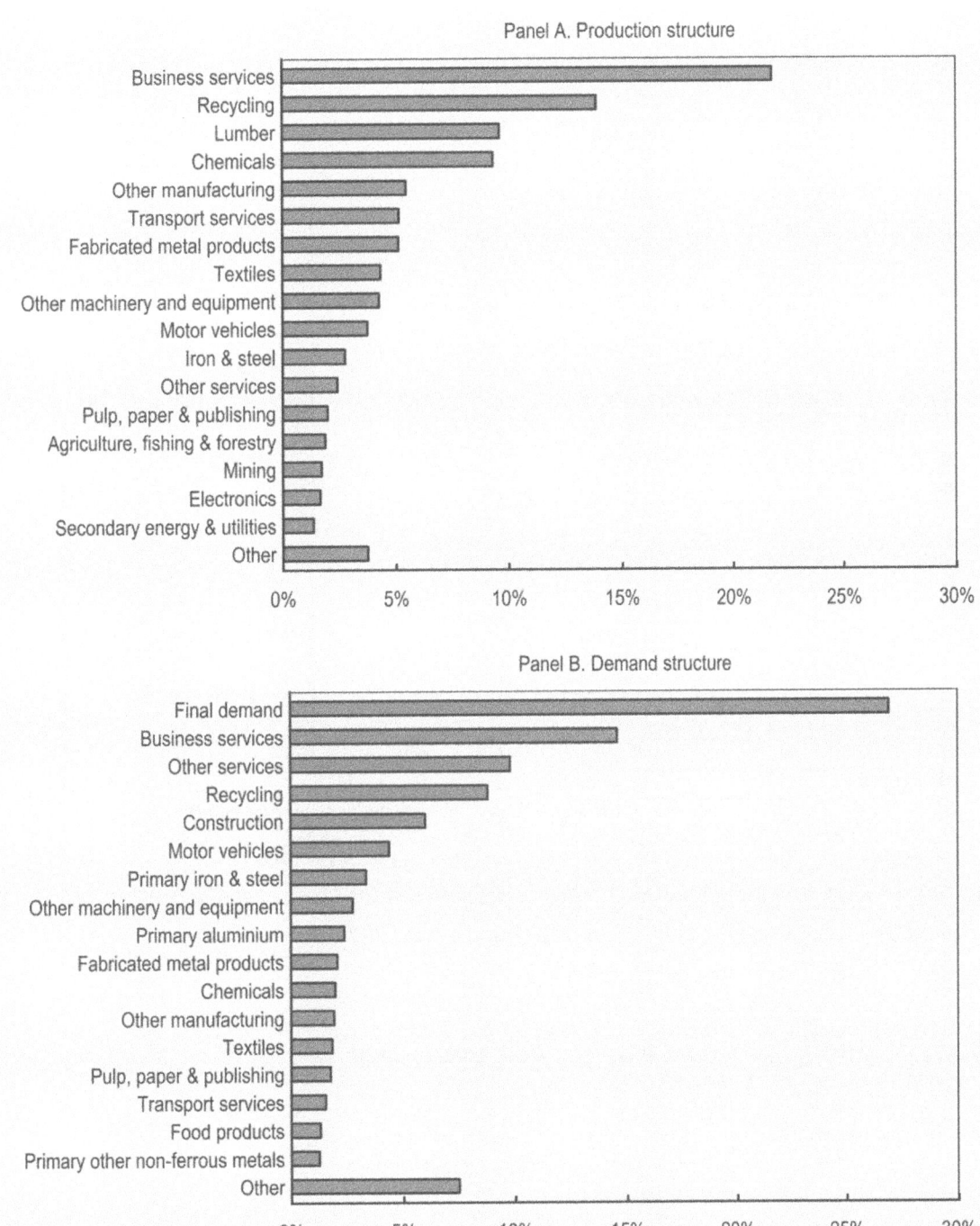

Source: Own compilation from Exiobase 3 database (Stadler et al., 2018[8]).

Figure 6.A.2. Production tax rates for recycling

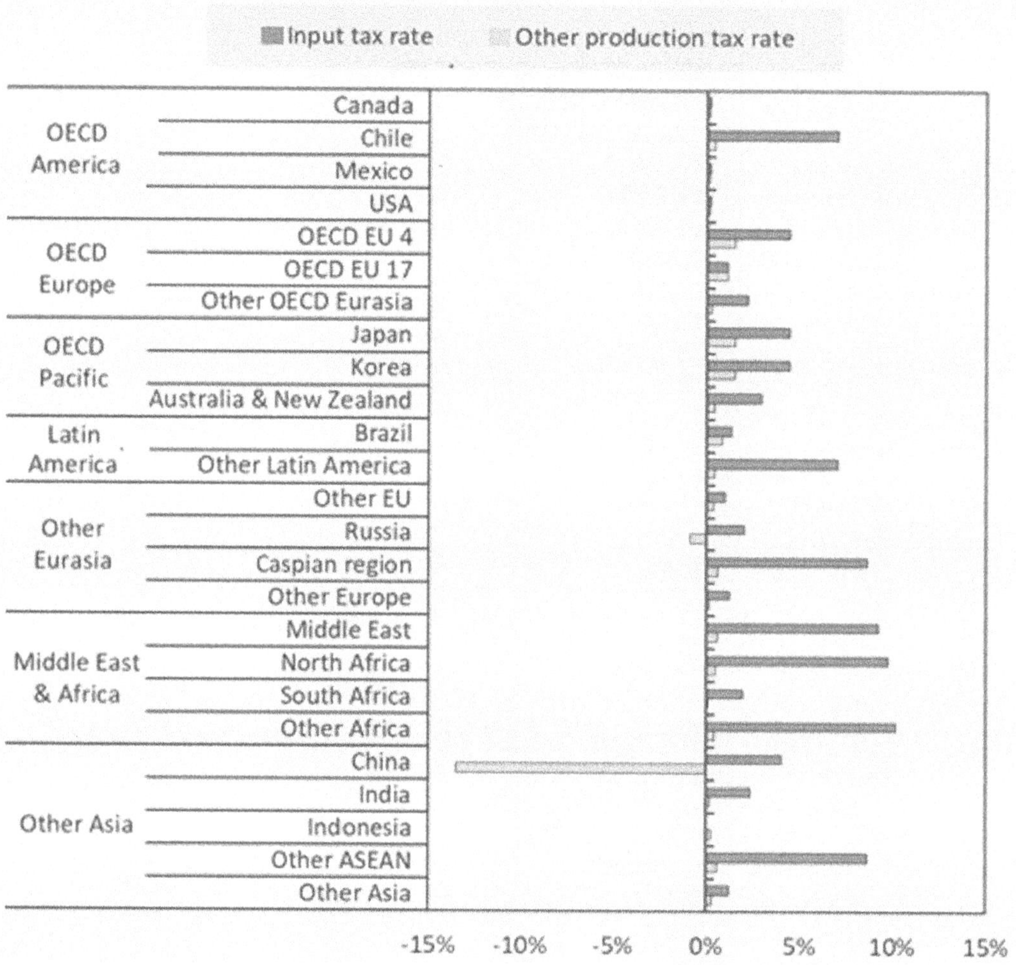

Source: Own compilation from Exiobase 3 database (Stadler et al., 2018[8]).

Chapter 7.

Case studies on demand and supply risks for specific materials: copper, iron and steel and critical materials

This report focuses on the big picture: global aggregated material flows. However, each material can be explored in more detail to provide insights into the economic mechanisms that are at play in the links between materials use and their economic drivers, and the prospects for supply risks. This chapter presents two case studies that model stock accumulation, waste generation and secondary materials use for copper and steel. The last section delves into the evolution of supply risks and the economic importance of materials for OECD countries to 2030 to assess potential criticality of materials in the future.

CHAPTER 7. CASE STUDIES ON DEMAND AND SUPPLY RISKS FOR SPECIFIC MATERIALS

KEY MESSAGES

This chapter adds some detail to the aggregate analysis of the economic drivers of materials use in the preceding chapters. It uses three detailed case studies to focus the spotlight on supply risks for specific materials that are critical for the OECD from both an economic and environmental perspective:

- A modelling of a scenario for future **copper** stocks and flows in Europe and China, and global projections. Copper is one of the most widely used metals in the economy, and copper extraction and processing has significant environmental impacts.

- A modelling of two scenarios for **iron and steel** to project the impacts of greater efficiency in steel-intensive industrial sectors, especially in China and India. Iron and steel production is the largest metals sector in volume, and its use is widespread across the economy.

- A modelling of two scenarios to project supply risks to **critical materials** by 2030, especially metals. Materials are considered to be critical when there is a significant risk of disruptions in supply and when supply disruptions have large economic impacts.

Projections and trends

Copper

- Global copper stocks are projected to more than double from current levels, to reach 1.1 Gt in 2050. China is projected to have by far the largest share of the 2050 global stock (approximately 40%, or 450 Mt of copper). Copper stocks in Europe and Japan are projected to stagnate by 2050, and growth in North America is projected to slow down.

- China's primary production from copper ores is triple that of Europe. While the share of secondary materials in making refined copper is equivalent (30%), Europe's secondary semi-finished goods production is much higher (21% vs. 5%), leading to a much higher content of recycled copper in end-products. The amount of scrap collected is 25% higher in Europe than in China. This reflects Europe's more mature equipment and infrastructure, as well as the higher rate of (end-of-life) scrap recycling (61% vs. 52%).

- A further decoupling of copper demand from economic development in emerging and developing economies is likely in the long run. While this theoretically reduces the need for primary production, it also limits the amount of scrap and secondary copper available in the future.

Iron and steel

- In the *Baseline* scenario global total steel production quadruples to 2060, with the *Increased Efficiency* scenario about 10% lower. This suggests that efficiency improvements can play a key role in reducing global demand. Most of the reduction is met through decreased primary steel production, and through increased secondary steel production.

CHAPTER 7. CASE STUDIES ON DEMAND AND SUPPLY RISKS FOR SPECIFIC MATERIALS | 163

- By 2060, the share of secondary production in total production is 48% in the *Baseline*, but reaches 53% in the *Increased Efficiency* scenario due to lower demand and the dynamics of scrap availability.

- Steel scrap production is projected to grow faster than iron ore production. Global scrap supply increases about 10 times by 2060 compared to 2011 in *Baseline* and around 8 times under the *Increased Efficiency* scenario. For China, this implies lower growth in iron ore demand under *Baseline*, and an absolute reduction in the *Increased Efficiency* scenario.

Critical materials

- Currently heavy rare earth elements (commonly used in car manufacturing, wind turbines, alloys and lighting) have the highest supply risk for the OECD countries.

- If production can shift to countries with large reserves, supply risks in the OECD tend to decrease for most metals in the model, except where reserves are concentrated in relatively few countries or in politically less stable countries. In addition, the low substitutability and low recycling rates of some materials magnify the supply risk.

Policy implications

An important consequence of the extensive use of non-renewable resources is the exhaustion of economically competitive domestic reserves, leading many industrialised economies to become increasingly dependent on the political stability of mineral-exporting countries.

These case studies confirm the assumption in the central baseline scenario that there is room for a significant scaling up of secondary production. However, shifts in the structure of the economy and economic decoupling may lead to the demand for metals growing less rapidly than GDP – limited demand growth can affect both primary and secondary metal production.

7.1. Detailed case studies bring global patterns into sharper focus

While the analysis of the economic drivers of materials use in the preceding chapters is necessarily performed in an aggregated form, this chapter uses three detailed case studies to put the central baseline scenario into perspective. More specific insights can be obtained by applying dedicated models that cannot cover the broad range of mechanisms outlined in Chapter 3, but that add further detail on specific aspects.

The case studies that investigate the supply risks associated with the increase in the use of primary and secondary materials. Each of these stand-alone analyses has been fully harmonised with the socioeconomic drivers presented in Chapter 3. They are thus complementary to the results presented in earlier chapters.

The first case study creates detailed projections of future copper flows to highlight and compare how stocks of copper build up over time in different countries, and how scrapping of copper provides a source of secondary materials. The second case study uses a model that is more similar to ENV-Linkages, but with a focus on the iron and steel sector. In both cases, the metals concerned are important from both an economic and environmental perspective. The case-studies allow a much more detailed representation of the technologies available for the penetration of secondary materials, and they allow for an accounting of metal stocks over time. This can provide crucial additional insights into supply risks for the provision of secondary metals, but also into the prospects for the saturation of metal stocks.

The third stand-alone analysis takes a somewhat different approach to evaluating supply risks and focuses on primary materials. It uses a fairly simple methodology to look at the criticality of specific materials in the OECD for the coming decades. This case study highlights which of the materials presented in this report are critical, but extends that by looking at projected developments of specific metals that are aggregated into groups in the central baseline scenario. While the information base does not allow for a full overview of the criticality of all materials globally to 2060, it illustrates how the baseline scenario can affect criticality – and thus supply risks – of various metals in the coming decades.

7.2. The copper case highlights the close links between primary and secondary material flows

Copper is one of the most widely used metals in the economy. It is used in energy production and transmission, in construction, water and heating pipes are often made of copper, and it is widely used in electrical wiring and electronics. Known copper reserves have increased in recent decades, and the estimated reserve life has been fairly stable at around 40 years (see Chapter 2). Furthermore, the properties of copper ensure that in theory it can be fully recycled, albeit not always in a cost-effective manner.

A detailed analysis on copper has been carried out with a dynamic material flow model (Glöser, Soulier and Tercero Espinoza, 2013[1]; Soulier et al., 2018[2]; Soulier et al., 2018[3]), using the ENV-Linkages central baseline scenario as a basis. The model portrays the entire value chain of copper from mining to recycling. Sixteen different end-use applications are distinguished. Global projections are complemented with regional projections from dedicated regional models for the European Union, North America, China, Japan and Latin America.

Figure 7.1 illustrates copper systems for 2015 for the European Union and China. These diagrams, known as Sankey diagrams, show the flows of copper through the domestic system; the width of the arrows reflect the size of the flows.

Figure 7.1. Copper material flows in 2015 differed significantly across countries

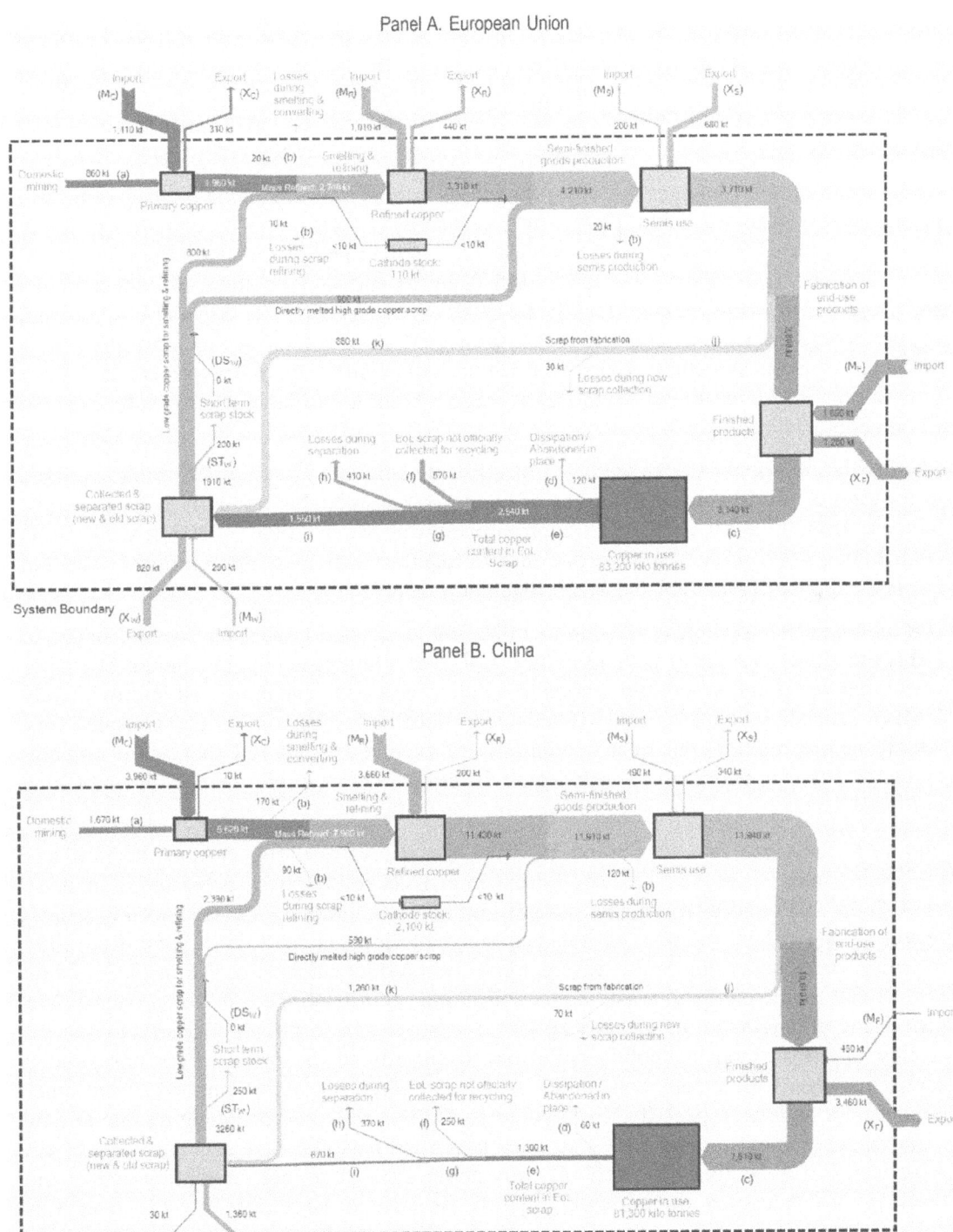

Source: Analysis by Fraunhofer ISI.

The two systems are similar for the size of the stocks of copper being used in the economy (slightly above 80 Mt), but not necessarily extracted domestically. However, China's primary production from copper ores is triple that of Europe. While the share of secondary materials in making refined copper is equivalent (30%), Europe's secondary semi-finished goods production is much higher (21% vs. 5%), leading to a much higher content of recycling copper in end-products. At the other end of the copper cycle, the amount of scrap collected is 25% higher in Europe than in China. This reflects Europe's more mature equipment and infrastructure, as well as the higher rate of (end-of-life) scrap recycling (61% vs. 52%). Furthermore, Chinese copper production – both in the form of semi-finished goods and final products – is geared towards exports, much more so than Europe.

Based on the ENV-Linkages sectoral projections, global anthropogenic copper stocks are projected to rise to 1.4 Gt in 2060 – an increase by a factor of 3.2 compared to 2015. China is projected to have by far the largest share in the 2060 global stock (approximately 40%, or 564 Mt of copper). Europe, North America and Japan's share of copper stocks are projected to decrease significantly by 2050 according to these calculations. Indeed, copper stocks in Europe and Japan are projected to stagnate, while slow growth is projected in North America. Apart from China, only Latin America is projected to have growth rates above or around the global average, so that it will have slightly surpassed the EU level in 2060.

The buildup of copper stocks leads to considerable amounts of scrap becoming available in the future. Worldwide, the amount of copper scrap available annually is projected to increase from 12.5 Mt in 2015 to approximately 43 Mt in 2060 (+250%), with China contributing 41%; the EU and North America together contribute approximately account for 16%, and Latin America and Japan together another 9%.

Not all copper scrap is recycled, due to collection losses and processing inefficiencies. In 2015, the global average of the end-of-life (EoL) recycling rate, i.e., the amount of copper waste that was transformed into secondary copper, was 43%, meaning that approximately 5 Mt of secondary copper is actually recycled. Since collection and processing rates vary widely across regions, the relative amounts of secondary material that are eventually extracted from copper scrap in each region differ (see the examples of Europe and China in Figure 7.1).

The EoL recycling rate in Europe in 2015 was 61%, resulting from collection and processing rates of around 80%. The European EoL recycling rate is projected to increase to 80 % by 2035 and then remain constant until 2050. China had a somewhat lower EoL recycling rate of 52% in 2015, but this is projected to increase to 70% in 2060. Recycling in China varies according to the different end-use product categories. For instance, while the official municipal solid waste collection is not well developed in China overall, the processing efficiency of the actually collected waste is comparatively high due to manual sorting operations run by mobile collectors, especially in the much larger unregulated scrap markets. China also constitutes a large market for scrap imports from other countries or regions, especially North America, though this may be changing due to new Chinese government regulations.[1] This is an important reason for the relatively low North American EoL recycling rate in 2015. Only about 25% of copper containing end-of-life products was recycled domestically, mainly because a large fraction of scrap was collected and eventually exported – mostly to China – but not officially registered (USITC, 2013[4]).

Figure 7.2. Copper stocks influence waste generation and recycling volumes

Copper stocks and flows in Mt

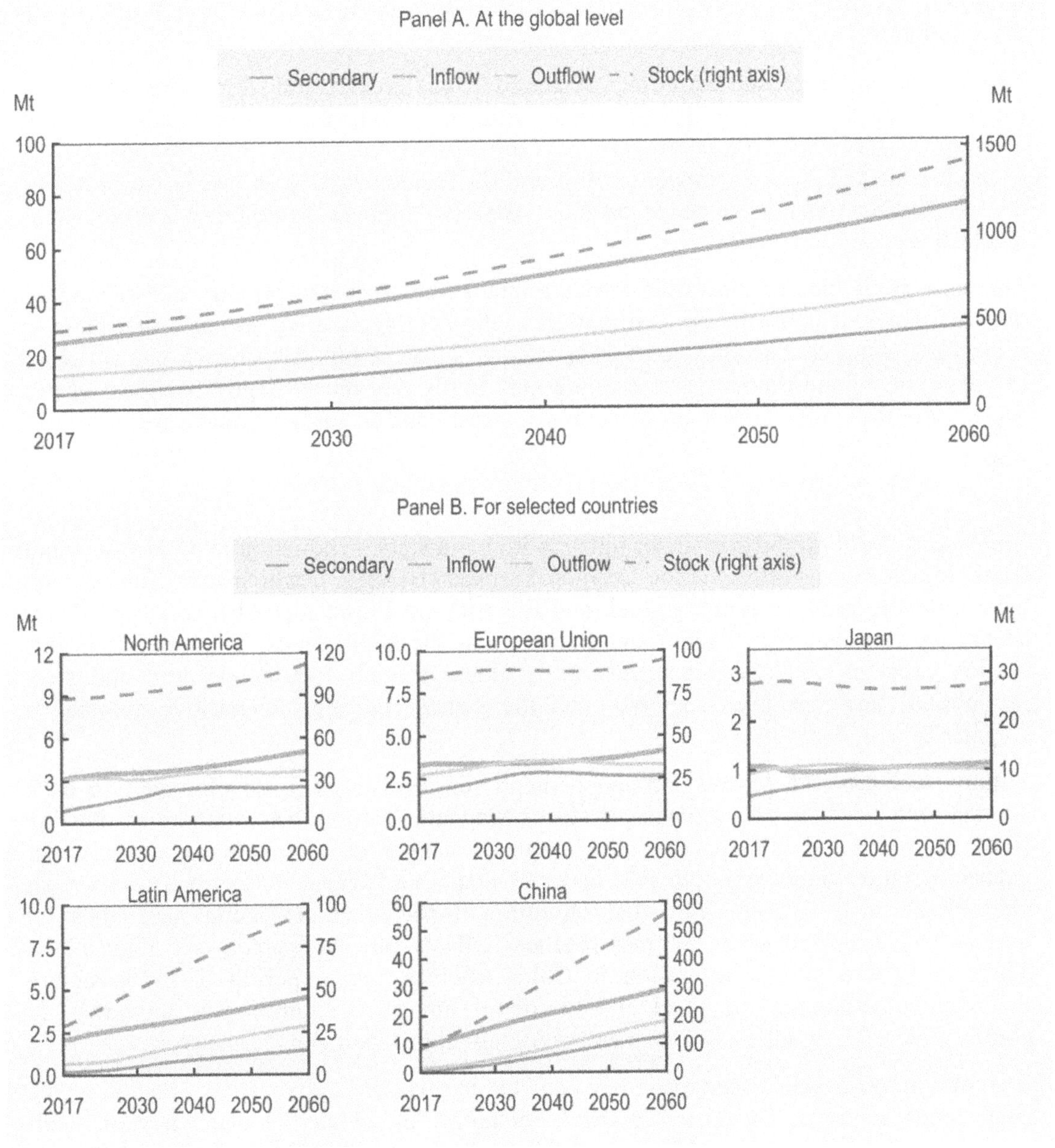

Panel A. At the global level

Panel B. For selected countries

Note: Simulated end-use copper stock (right axis), yearly inflows (production) and outflows (scrapping) of copper contained in end-use products, and yearly amount of recycled secondary material (left axis) in Mt.
Source: Analysis by Fraunhofer ISI based on the central baseline scenario of the OECD ENV-Linkages model.

StatLink ⇒ https://doi.org/10.1787/888933885429

In the projections presented here, copper demand is proportionally tied to sectoral economic development. Historically, this has been a reasonable assumption. However, for China, projected economic development in combination with an expected population decline would lead to a per-capita copper stock of approximately 420 kg in 2060. In comparison, per-capita stocks of some OECD economies show saturation effects as they have stagnated at around 200 to 250 kg. It is therefore likely that at global level a further decoupling between economic development and copper demand will take place in the long run.[2] While this theoretically reduces the need for primary production, it also limits the amount of scrap and secondary copper becoming available in the future.

The global and regional evolution of the relevant stocks and flows are shown in Figure 7.2. At the global level, total production, scrapping, recycled secondary material and copper stocks all grow at comparable speeds (panel A). This is driven by fast growth in production and stock accumulation in non-OECD countries, as shown in panel B. For the OECD countries represented in panel B, secondary material availability is projected to level off after around 2040.

This case study clearly shows the intricate interactions between primary and secondary material flows. It confirms the assumption in the central baseline scenario that there is room for a significant scaling up of secondary production, but that shifts in the structure of the economy and economic decoupling may imply that the demand for copper grows less rapidly than GDP, which can affect both primary and secondary copper use.

7.3. The case of iron and steel reveals very large recycling potential

The largest metal production sector today is iron and steel. Production involves two main technologies: Blast-Furnace/Basic Oxygen Furnace (BF-BOF), which uses iron ore and coke inputs to produce primary steel, and Electric Arc Furnace (EAF), which produces secondary steel using steel scrap and electricity as the main inputs. The analysis in this section uses the ENGAGE-materials CGE model (which focuses on iron and steel production) and has been calibrated to the central baseline scenario presented in Chapter 2.[3]

Future steel demand depends on the demand for infrastructure development in each country and whether per-capita steel stocks stagnate (saturate) as economies develop (Pauliuk, Wang and Müller, 2013[5]). The rate of steel stock accumulation and turnover determines the amount of scrap that becomes available for secondary steel production. Bleischwitz et al. (2018[6]) find that saturation in the USA per-capita steel stock (at around 16 t) occurred when incomes reached USD 16 000 per person (see Figure 7.3). There may be a similar saturation in China after its recent period of infrastructure development, although the level and timing of any such saturation is unknown. In contrast, there are no clear signs of a saturation effect in Japan and Germany.

Two alternative baseline scenarios are calculated in this case study: (i) *Baseline*, in which past trends continue, and (ii) *Increased Efficiency*, in which the efficiency of steel-intensive industrial sectors in China and India converges with those in the United States. This improved efficiency concerns the input of steel to the largest steel-consuming sectors (construction, motor vehicles, metal products and machinery). As stressed in Chapter 2, these baseline scenarios assume no new policies. The structural changes outlined in Section 3.2 limit growth of steel demand to rates below GDP growth, but there is no explicit assumption of "peak demand" in these projections.[4] Steel production is therefore projected to be substantially higher than long-term projections of demand made by the World Steel Association and the projections discussed within the context of the OECD Steel Committee that include plausible policy trends.

Figure 7.3. Per-capita stocks of steel have stabilised in some developed countries

10 year moving averages of per-capita steel stocks

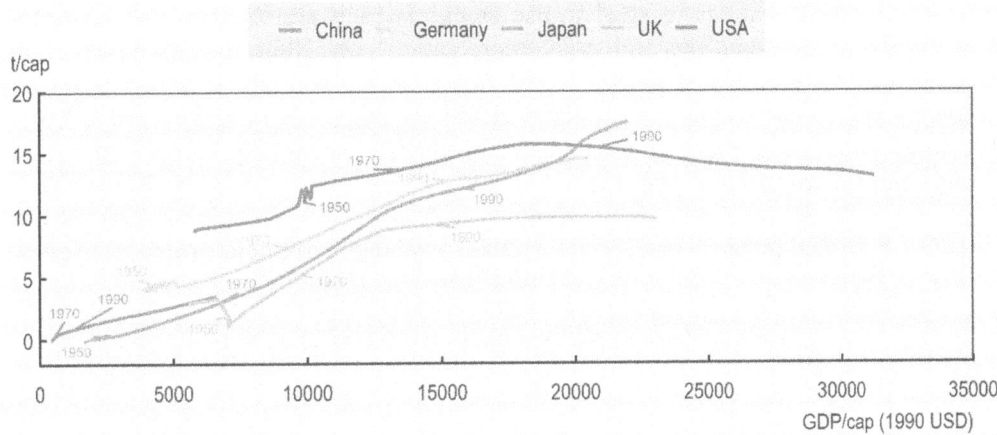

Source: Adapted from Bleischwitz et al. (2018[6]).

This is highlighted in e.g. (OECD, 2018[7]): "Over the longer term, factors such as circular economy […] could weigh on steel demand growth." Furthermore, most existing analyses concentrate on OECD countries, plus China and India, while Section 3.1 clearly shows that the surge in growth in developing economies, not least in Sub-Saharan Africa, may bring a significant boost in materials demand, including steel.

In the Baseline scenario, global total production of steel roughly doubles to 2060 (Figure 7.4).

Figure 7.4. Global steel production is projected to grow significantly

Global steel production in tln USD (2011 PPP)

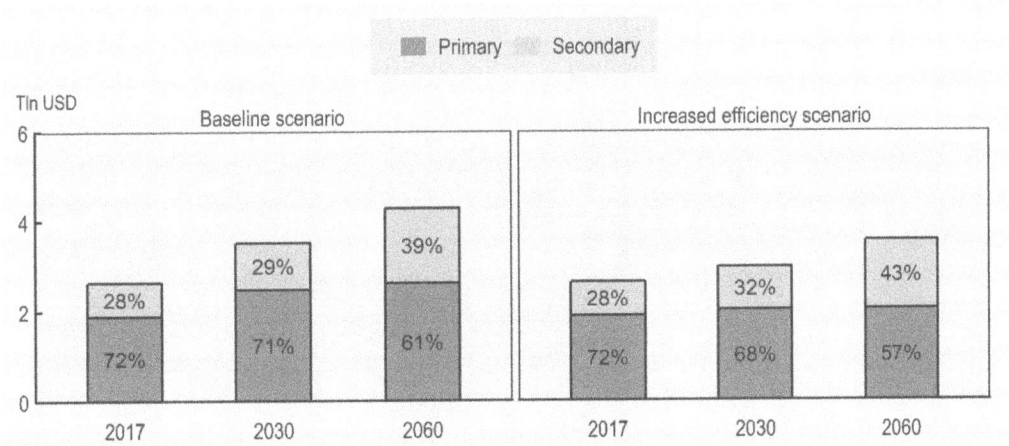

Source: UCL's ENGAGE model, based on the central baseline scenario of the OECD ENV-Linkages model.

StatLink https://doi.org/10.1787/888933885467

The *Increased Efficiency* scenario is about 17% lower than *Baseline* in 2060. The associated iron ore inputs increase from less than 3 Gt to more than 6 Gt. This suggests that efficiency improvements can play a key role in reducing global demand. The majority of the reduction is met through decreased primary steel production.

Annual scrap availability for secondary steel production in each region is calculated by taking into account changes in steel in-use stocks (which are themselves updated with new steel production quantities and stock depreciation). By 2060, the share of secondary production in total production is 39% in the *Baseline*, but reaches 43% in the *Increased Efficiency* scenario due to lower steel demand and the dynamics of scrap availability. The ability to model both primary and secondary production separately captures the internal dynamics and substitutions at play here.

Figure 7.5. Global primary and secondary steel production are both projected to grow significantly

Panel A. Regional steel production (tln USD in 2011 PPP)

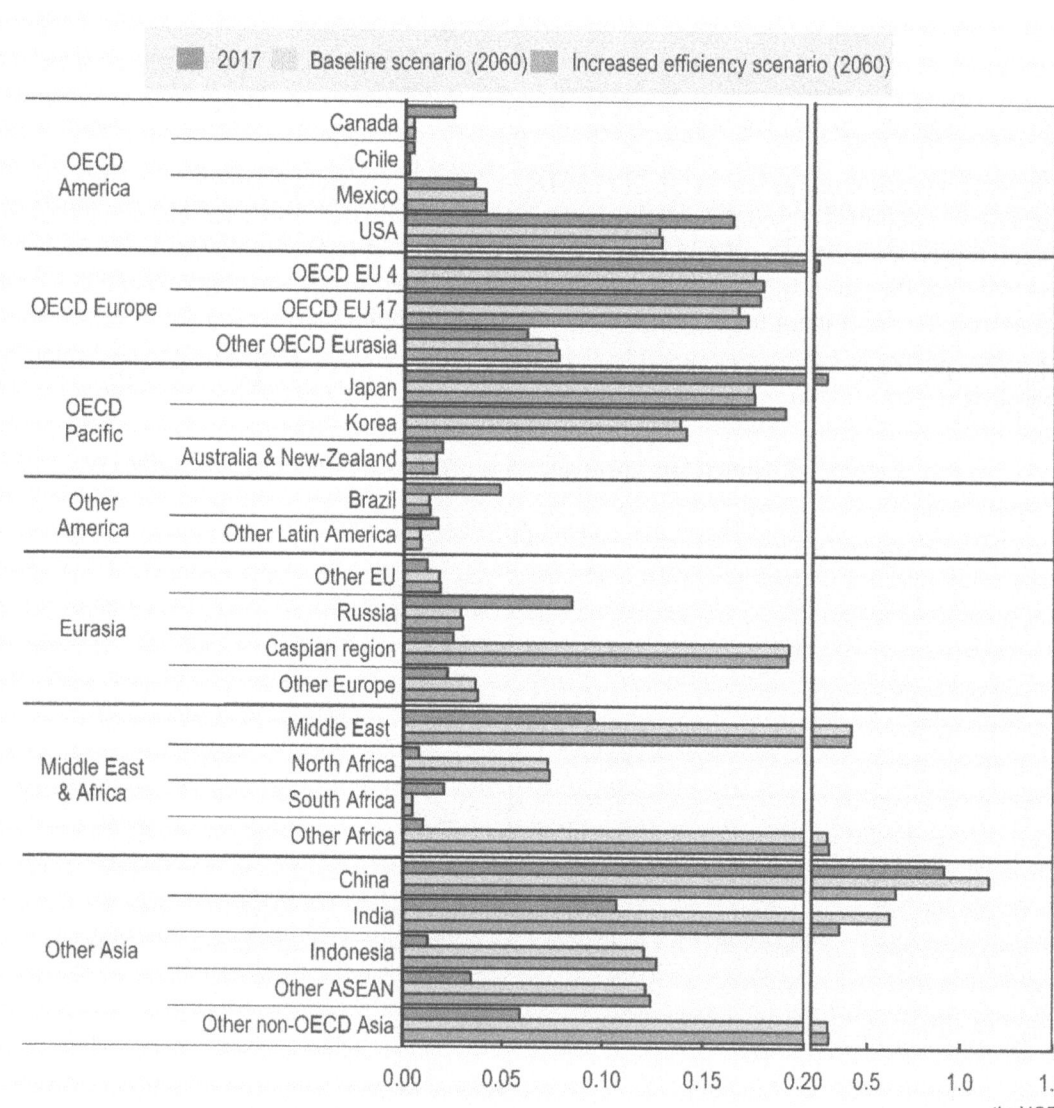

Figure 7.5. Global primary and secondary steel production are both projected to grow significantly (*continued*)

Panel B. Shares in total steel production in selected countries

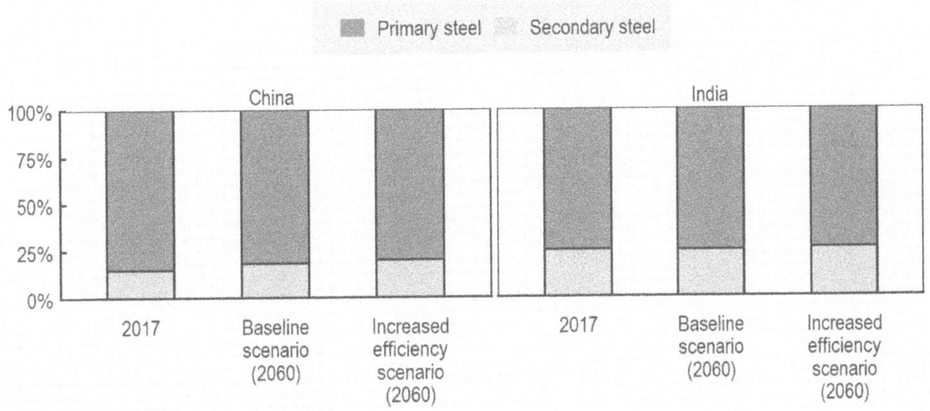

Note: See Table 2.1 for regional definitions. In particular, OECD EU 4 includes France, Germany, Italy and the United Kingdom. OECD EU 17 includes the other 17 OECD EU member states. Other OECD Eurasia includes the EFTA countries as well as Israel and Turkey. Other EU includes EU member states that are not OECD members. Other Europe includes non-OECD, non-EU European countries excluding Russia. Other Africa includes all of Sub-Saharan Africa excluding South Africa. Other non-OECD Asia includes non-OECD Asian countries excluding China, India, ASEAN and Caspian countries.
Source: UCL's ENGAGE model, based on the central baseline scenario of the OECD ENV-Linkages model.

StatLink https://doi.org/10.1787/888933885486

By 2060, *Baseline* scenario steel production significantly increases in most countries; in particular it doubles in China and grows even more significantly in India (Figure 7.5, panel A). Introducing the *Increased Efficiency* scenario assumptions leads to lower levels of total steel production in China and India, while some other regions increase their steel output marginally in response as the decrease in steel demand in these two major countries leads to a reduction in the global steel price. An important finding is that the share of secondary steel production in China increases by 2060 in the *Increased Efficiency* scenario, due to increased scrap availability and lower regional steel demand. In India, the share of secondary steel production remains more stable due to a higher share of scrap already in 2011 and an increased production of both scrap and iron ore (Figure 7.5, panel B).

These steel production trajectories lead to steel stocks in China of 40 t per capita in 2060 in the *Baseline* scenario and 26 t per capita in the *Increased Efficiency* scenario (Figure 7.6). These estimates are influenced by a declining population after 2030, which dampens the decline in per-capita stocks. The change in the structure of the economy in the Baseline scenario would thus lead to levels of stocks per capita that are higher than those in current industrialised economies, which questions any projections that exclude the development of efficiency of steel use in downstream sectors.

Per-capita stocks in India start at a low level and the two scenarios only begin to diverge from the late 2030s onwards (Figure 7.6). Even with the significant growth in steel production, stocks only reach about 5.5 tonnes per capita by 2060 in *Increased Efficiency* and 8 tonnes per capita in *Baseline*. Therefore, India appears to be accumulating less steel than China at similar levels of GDP per-capita. This reflects a lower steel intensity of capital goods and infrastructure in India.

Figure 7.6. Growth in per-capita steel stocks is projected to continue in China and India

Steel stocks per capita in t/cap.

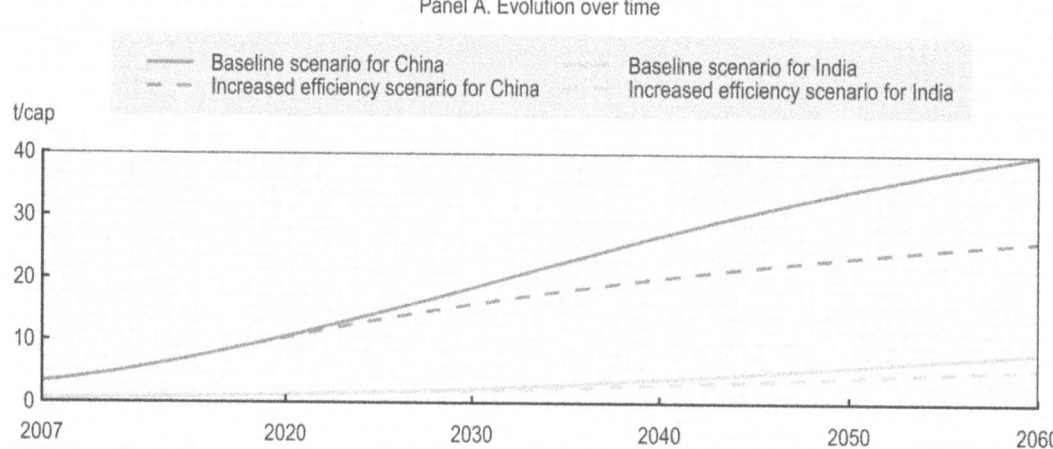

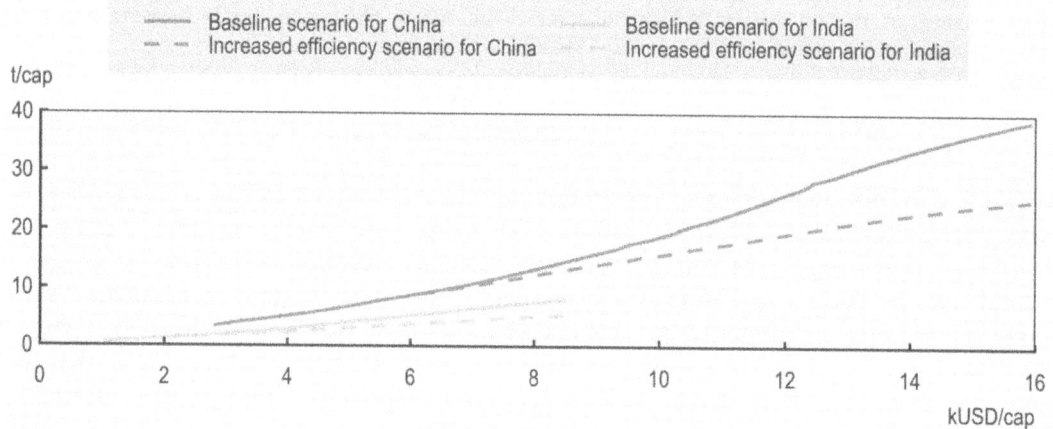

Source: UCL's ENGAGE model, based on the central baseline scenario of the OECD ENV-Linkages model.

StatLink https://doi.org/10.1787/888933885505

Global scrap supply increases roughly 3 times (i.e. increases by around 200%) by 2060 compared to 2017 in *Baseline* and around 2.7 times under the *Increased Efficiency* scenario (Figure 7.7, panel A). This increase in scrap drives a gradual substitution of primary steel by secondary steel, with a reductive effect on global iron ore demand. For China, a country that uses all its iron ore domestically, a higher availability of scrap relieves pressure on iron ore production, implying a lower growth compared to that of steel production under *Baseline* scenario assumptions (Figure 7.7, panel B). A lower availability of scrap in the *Increased Efficiency* scenario slightly reverses that situation. For iron ore exporting countries such as India, the projected reduction in domestic ore demand – because of the 9 times increase in scrap supply in the country – is offset by an increase in iron exports to other regions (Figure 7.7, panel C).

Figure 7.7. Steel scrap production is projected to grow faster than iron ore production

Cumulative growth between 2017 and 2060

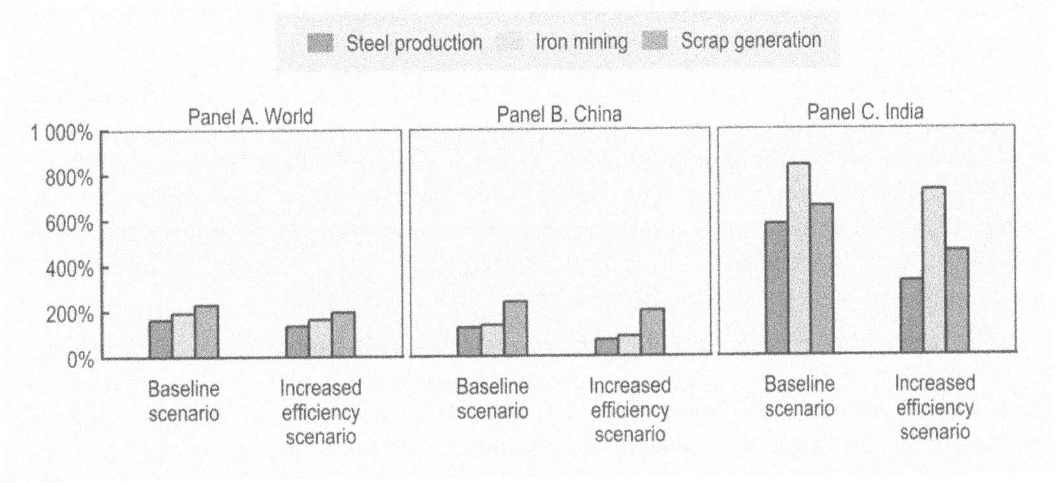

Source: UCL's ENGAGE model, based on the central baseline scenario of the OECD ENV-Linkages model.

StatLink https://doi.org/10.1787/888933885524

The more general insights from this detailed analysis are fairly similar to the analysis of copper flows: at least at the global level secondary production is projected to grow more rapidly than primary materials use, and in case of low or even negative growth in demand for the processed commodities, primary materials use may potentially even decline over time in baseline scenarios.

7.4. The case of critical materials in the OECD shows the key role of supply risks

The primary materials assessed in this report are non-renewable resources (apart from biotic materials). Moreover, their deposits in the Earth's crust are often geographically clustered, which makes the security of supply a potential risk for resource-importing countries. An important consequence of the extensive use of non-renewable resources is the exhaustion of economically competitive domestic reserves, which has made many industrialised economies increasingly dependent on the political stability of mineral-rich countries. This may be problematic especially when the over-exploitation of a mineral coincides with a high demand for economic production and low substitutability and recycling rates. Several reports have thus been developed to assess the vulnerability of economies to supply disruptions of minerals (European Commission, 2014[8]; European Commission, 2010[9]).

In this context, "critical minerals" refers to the group of non-renewable materials for which "the risk of disruptions in supply is relatively high and for which supply disruptions will be associated with large economic impacts" (Coulomb et al., 2015[10]). Indices for supply risk and economic importance have been computed using the methodologies used by Coulomb et al. (2015[10]) and the European Commission (2014[8]). The supply risk index incorporates material substitutability, recycling rates and the concentration of production in countries that are judged by international datasets to be relatively politically unstable. Economic importance (i.e. vulnerability to supply risk) is computed based on in how much of the material is used in the different sectors and how economically important these sectors are for the overall economy.[5] Supply risk is presented on a scale of 0-5 and economic importance on a scale of 0-1. In this index calculation, materials with a supply risk above an index value of 1, and economic importance above 0.05 were deemed critical. These thresholds are the results of expert judgement.[6]

Assessing the criticality of the materials in the central baseline scenario of this report is challenging, since many of the materials, which were identified as most critical by previous studies (e.g. Coulomb et al. (2015[10])) are grouped under "Other metal ores" in the ENV-Linkages model. The criticality of specific materials within this group varies widely. Coulomb et al. (2015[10]) take a more granular approach and apply a quantitative method to assess the criticality of 51 specific materials. The assessment is limited to the OECD only, and does not extend beyond 2030.[7]

Table 7.1 shows the current criticality of the materials that were assessed in Coulomb et al. (2015[10]). According to this computation, critical materials include chromium ores, some other metal ores (rare earth elements were assessed with the highest supply risk; tungsten with the highest economic importance), platinum group metal ores, some fertiliser minerals (especially phosphate rock), a sub-group of chemical minerals (fluorspar with the highest supply risk, barytes with the highest economic importance) and some industrial minerals (natural graphite with the highest supply risk; magnesite with the highest economic importance). All selected materials in the table were deemed of high critical economic importance, except gold ores and some industrial minerals.[8]

Table 7.1. Criticality of selected materials in the OECD in 2012

Supply risk expressed on a scale of 0-5; Economic importance on a scale of 0-1

	Supply risk	Economic importance
Bauxite and other aluminium ores - gross ore	0.54	**0.09**
Chemical minerals (*)	**1.14 - 2.07**	**0.07 - 0.11**
Chromium ores	**1.06**	**0.10**
Coking Coal	0.96	**0.10**
Copper ores	0.22	**0.06**
Fertiliser minerals	0.17 - **1.38**	**0.07 - 0.09**
Gold ores	0.16	0.04
Gypsum	0.48	**0.05**
Industrial minerals (*)	0.28 - **3.24**	0.03 - **0.09**
Iron ores	0.90	**0.08**
Limestone	0.37	**0.06**
Manganese ores	0.67	**0.08**
Nickel ores	0.23	**0.09**
Other metal ores (*)	0.21 - **4.61**	**0.05 - 0.10**
Platinum group metal ores	**1.13**	**0.07**
Silver ores	0.30	**0.05**
Tin ores	0.88	**0.07**
Titanium ores	0.10	**0.07**
Zinc ores	0.58	**0.09**

Note: (*) indicates that not all materials in this group are included in the range; bold values denote critical levels.
Source: Own compilation based on Coulomb et al. (2015[10]).

StatLink https://doi.org/10.1787/888933885885

The criticality of materials may change over time, following the evolution of supply risk and economic importance. However, projecting future supply risks in a robust way is challenging, as they depend on future geopolitical stability (among others). As in Coulomb et al. (2015[10]), this projection assumes unchanged geopolitical stability. Supply risk thus varies over time with the concentration of production in countries with different degrees of geopolitical stability.

The central baseline scenario presented in this report is used to update the assessment of economic importance in 2030. For projecting future supply risk, Coulomb et al. (2015[10]) presented two scenarios. In the *Constant production shares* scenario, production shares of materials across regions stay constant over time. As a consequence, supply risks remain constant over time in this scenario. In the Shifting *production shares* scenario, materials production gradually moves towards regions with high known reserves. Figure 7.8 presents an update of the criticality of materials in the OECD by 2030 for both scenarios, using the central baseline scenario of this report as input.

The *Constant production shares* scenario projects only small changes between 2012 and 2030. In this scenario (presented in Figure 7.8, panel A), supply risks stay unchanged since the global distribution of production as well as the geopolitical stability of individual countries are assumed to be constant over time. Furthermore, the economic structure of the OECD is projected to remain similar to the current one, which implies stable values for the economic importance indicator.

In contrast, the criticality of several minerals is projected to decrease in the *Shifting production shares* scenario, as shown in panel B of Figure 7.8. As the global distribution of mineral production shifts towards countries where known reserves are high, the supply risk of several minerals decreases significantly, hence the reduced criticality.

Currently, and in the *Constant production shares* projection, heavy rare earth elements (commonly used in car manufacturing, wind turbines, alloys and lighting), have the highest supply risk. However, the supply risk is driven by current production shares. If the geographical spread of production becomes better aligned with reserves (as shown in panel B) then these elements will become less critical. Germanium, a semi-conductor commonly used to make electronic devices, as well as light rare earth elements are similarly projected to see a decline in supply risk. Provided that production can shift towards regions with large reserves, they would lose their criticality status.

Although shifting production shares results in a decline in supply risk for most minerals, it can also have the opposite effect for some minerals. For instance, Antimony, which is used for the production of batteries and glasses, is considered critical in both cases, but projected to become the material with the highest supply risk and criticality in the *Shifting production shares* scenario. Similarly, iron ore is projected to become critical in the *Shifting production shares* scenario.

The increase in supply risk for different materials comes from the fact that production shifts to countries with the largest reserves – but these countries are either less politically stable or capture a large share of the global market. For instance, China is projected to dominate the supply of antimony, fluorspar, gallium, germanium, graphite, indium, magnesium, rare earths, and tungsten, while Russia is projected to be the dominant supplier of platinum group metals. In addition, the low substitutability and low recycling rates of these materials magnify the supply risk.

176 | CHAPTER 7. CASE STUDIES ON DEMAND AND SUPPLY RISKS FOR SPECIFIC MATERIALS

Figure 7.8. If production shifts towards countries with large reserves, only few materials remain critical in the OECD by 2030

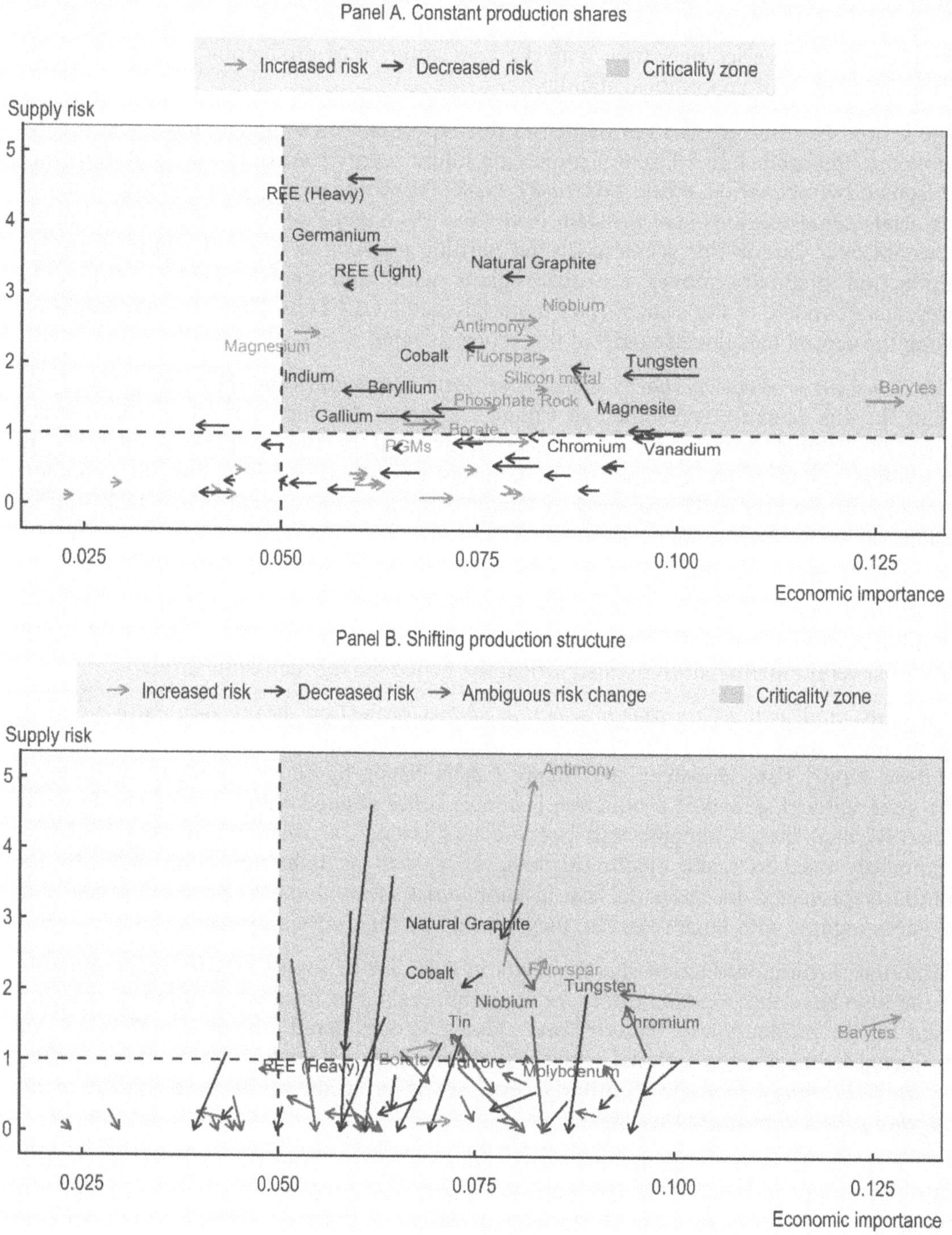

Note: The red zone represents the criticality, defined as the supply risk above 1 and the economic importance above 0.05. The chart only displays the names of the materials projected to remain or become critical.
Source: Own calculations based on Coulomb et al. (2015[10]).

StatLink https://doi.org/10.1787/888933885543

Barytes, used for mud drilling in the oil and gas sectors, are projected to get the highest score for economic importance in 2030 in both scenarios. This is due to a projected increase in the share of OECD countries in producing oil (and gas), which implies an increase in economic importance for the OECD region. For all other materials, changes in economic importance are much smaller, which is intuitive given the relatively stable economic development of the OECD economies.

Looking at longer time horizons, and taking a global perspective, criticality assessments may start to shift much more profoundly. This assessment is limited to OECD countries and a 2030 time horizon, since predicting the evolution of the geopolitical situation of individual countries (in particular outside the OECD) over a longer time frame is a challenging exercise and outside the scope of this report.

Notes

[1] See for instance notifications G/TBT/N/CHN/1211 and G/TBT/N/CHN/1212 by the Chinese government to the Committee on Technical Barriers to Trade of the World Trade Organization.

[2] Such relative decoupling is also present in the central baseline scenario presented in the other Chapters of this report; see Section 5.3.

[3] Data from EXIOBASE and the World Steel Association are used and reconciled to specify the technology cost structures of both primary and secondary steel sectors.

[4] The discussion on whether a peak in global steel demand can be expected before 2060 is ongoing, and studies exist to support both sides (OECD Steel Committee, 2018). Given the long-term nature of steel production plants, this discussion has important consequences for the question of whether existing "excess capacity" will disappear because of increasing demand or remain.

[5] See Coulomb et al. (2015[10]) for detailed methodology and formulae.

[6] As discussed in European Commission (2010[9]): "The thresholds used to distinguish high from lower supply and environmental risks or economic importance have been determined pragmatically and inevitably involve a certain judgment as there is no unequivocal methodology in this domain. [...]. It should be stressed that the distinction between "critical" raw materials and other raw materials is the result of a relative, rather than an absolute, assessment and that the quantitative methodology not only restricts inevitably the number of factors that can be taken into consideration but also that this assessment provides only a static view of the situation. In particular, it is important to note that the supply risks for some raw materials can change relatively rapidly."

[7] Coulomb et al. argue that, extending the analysis beyond 2030 is problematic as the uncertainties regarding key factors, such as political risk associated with supply countries, become too large. The update conducted for this report thus adopts the same time frame and regional scope.

[8] Certain materials were excluded from the assessment by Coulomb et al. (2015[10]). Exclusion was based on an *a priori* expert judgement, which considered the supply risk of such materials too low (e.g. Lead ores, Sand and gravel).

References

Bleischwitz, R. et al. (2018), "Extrapolation or saturation – Revisiting growth patterns, development stages and decoupling", *Global Environmental Change*, Vol. 48, pp. 86-96, http://dx.doi.org/10.1016/J.GLOENVCHA.2017.11.008. [6]

Coulomb, R. et al. (2015), "Critical minerals today and in 2030: an analysis for OECD countries", *Environment Working Paper No. 91*, http://www.oecd.org/environment/workingpapers.htm (accessed on 10 January 2018). [10]

European Commission (2014), *Report on critical raw materials for the EU*, European Commission, Brussels. [8]

European Commission (2010), *Defining critical raw materials Critical raw materials for the EU*, European Commission, Brussels, http://ec.europa.eu/enterprise/policies/raw-materials/documents/index_en.htm (accessed on 20 July 2018). [9]

Glöser, S., M. Soulier and L. Tercero Espinoza (2013), "Dynamic Analysis of Global Copper Flows. Global Stocks, Postconsumer Material Flows, Recycling Indicators, and Uncertainty Evaluation", *Environmental Science & Technology*, Vol. 47/12, pp. 6564-6572, http://dx.doi.org/10.1021/es400069b. [1]

OECD (2018), *Steel market developments*. [7]

Pauliuk, S., T. Wang and D. Müller (2013), "Steel all over the world: Estimating in-use stocks of iron for 200 countries", *Resources, Conservation and Recycling*, Vol. 71, pp. 22-30, http://dx.doi.org/10.1016/J.RESCONREC.2012.11.008. [5]

Soulier, M. et al. (2018), "Dynamic analysis of European copper flows", *Resources, Conservation and Recycling*, Vol. 129, pp. 143-152, http://dx.doi.org/10.1016/J.RESCONREC.2017.10.013. [2]

Soulier, M. et al. (2018), "The Chinese Copper Cycle: Tracing Copper through the Economy with Dynamic Substance Flow and Input-Output Analysis", *Journal of Cleaner Production*, Vol. Forthcoming. [3]

USITC (2013), "Executive Briefings on Trade Share Billion US$ Figure 3: China's Imports of Iron Ores/Concentrates", https://www.usitc.gov/publications/332/2013-04_China-Africa%28GamacheHammerJones%29.pdf (accessed on 17 May 2018). [4]

Part III.

The environmental consequences of materials use

Chapter 8.

Projections of the environmental consequences of materials use

> *This chapter provides insights into the environmental consequences of the materials use projections presented in the earlier chapters. It starts with a qualitative overview, and then presents projections of the impacts on climate change. It also includes an analysis of the broader environmental impacts of selected materials at the global level: first for 7 metals and then for concrete. This highlights some of the important impacts that materials use could have on the environment in the future.*

KEY MESSAGES

The extraction, processing, use and disposal of materials lead to a range of environmental consequences, including on climate change, pollution of land, water and air, as well as damages to ecosystems and biodiversity.

Projections and trends

- The environmental consequences of extracting, processing and using different material resources vary widely across material groups and the stage of the material life cycle. Toxicity and air pollution consequences are particularly large for metals extraction and processing. Fossil fuel use is closely associated with emissions that contribute to climate change and air pollution. Biomass production has a strong link with land use and water pollution. Non-metallic minerals extraction, processing and use have a more diverse set of impacts. For instance, fertiliser use leads to water pollution. Construction activities are linked to high greenhouse gas emissions.
- Global GHG emissions are projected to reach 50 Gigatonnes CO_2-equivalent (Gt CO_2-eq.) before 2030, and continue rising to around 75 Gt CO_2-eq. by 2060. The ambitions of the Paris Climate Agreement, including the Nationally Determined Contributions (NDCs) and the "well below two degrees" objective will thus not be met under the central baseline scenario. Emissions related to materials management (from the combustion of fossil fuels for energy, from agriculture, from manufacturing, and from construction activities) are currently the main sources of GHG emissions (about two thirds of total emissions).
- Emissions related to materials management are projected to increase in the coming decades from 30 Gt CO_2-eq. to almost 50 Gt CO_2-eq., despite their share in overall emissions decreasing slightly compared to emissions from transport, households and services (see Figure 8.1).
- Environmental impact assessment for seven key metals (iron, aluminium, copper, zinc, lead, nickel and manganese) highlights the wide range of environmental consequences linked to materials extraction, processing and use. These impacts include significant impacts on acidification, climate change, cumulative energy demand, eutrophication, human toxicity, land use, photochemical oxidation, and aquatic and terrestrial ecotoxicity. In general, copper and nickel cause the greatest per-kilo environmental impacts of the metals investigated in this report, while iron has the highest absolute environmental impacts due to the large volumes used.
- While the production, processing and use of secondary metals also lead to environmental impacts, these are generally an order of magnitude lower than that of primary production.
- The total environmental impacts of using these metals are projected to more than double and in some cases even quadruple by 2060.
- In comparison to metals, concrete has much smaller impacts per kilogramme. However, the volume of concrete used is much larger and thus its total environmental consequences are also quite significant. Concrete is especially associated with consequences for climate change. The seven metals and concrete together represent almost a quarter of all GHG emissions and one sixth of cumulative energy demand.

Figure 8.1. Emissions from materials management are projected to increase, but not more rapidly than other emission sources

Global GHG emissions in Gt CO_2-eq.

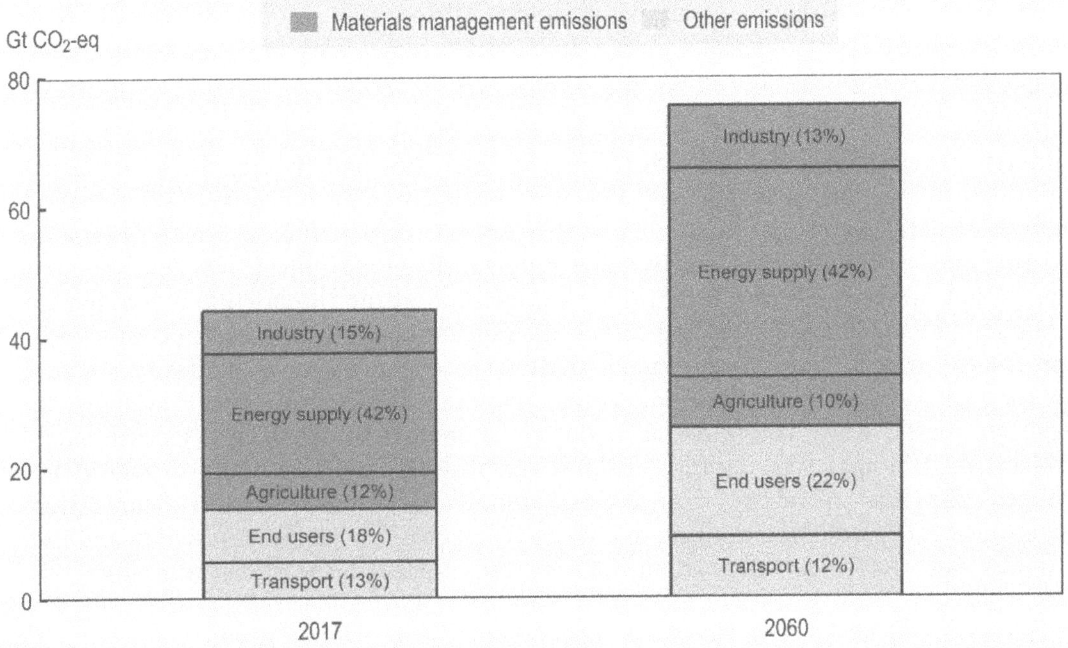

[Stacked bar chart showing Gt CO_2-eq for 2017 and 2060, differentiating Materials management emissions and Other emissions.

2017 (~45 Gt CO_2-eq):
- Industry (15%)
- Energy supply (42%)
- Agriculture (12%)
- End users (18%)
- Transport (13%)

2060 (~75 Gt CO_2-eq):
- Industry (13%)
- Energy supply (42%)
- Agriculture (10%)
- End users (22%)
- Transport (12%)]

Note: Emissions are differentiated between those linked to materials management and those that are not, based on previous OECD work on materials management within OECD countries (OECD, 2012[1]). For example, fossil fuel combustion by households is not attributed to materials management, while all industrial activity is. End users includes services and households as well as private transportation (light duty vehicles).
Source: OECD ENV-Linkages model.

StatLink https://doi.org/10.1787/888933885562

Policy implications

With such a wide range of environmental consequences linked to materials use, governments face the complex challenge of designing policy packages for improving resource efficiency and stimulating the transition to a circular economy while addressing the relevant trade-offs among different environmental issues. A well-designed policy package could, however, not only reduce materials use, but also lead to multiple environmental benefits.

Policies aimed at reducing emissions of GHGs should be included in a comprehensive resource efficiency policy package, as fossil fuel use is closely linked to GHG emissions and materials management activities form a large share of total GHG emissions. These links between resource efficiency and climate change go in both directions: a transition to a low-carbon economy is a key part of a resource-efficient, circular economy, and improving resource efficiency is a cornerstone of an ambitious climate mitigation policy package. There are, however, also trade-offs to consider, for instance between recyclability and energy efficiency in the construction of vehicles.

8.1. Materials use has many environmental consequences

One of the main motivations for improving resource efficiency and reducing materials use is to limit the environmental impacts that are linked to the use of the materials. The economic activities that drive materials use have a range of environmental consequences. Some of these consequences can be attributed directly to resource provision (e.g. greenhouse gas (GHG) emissions from extraction and processing of primary materials), while others are indirectly linked to resource use (e.g. air pollution caused by combustion of fossil fuels).

Different material resources have different characteristics and the activities associated with their extraction, management and use have different potential environmental implications (Table 8.1). Ore mining can cause air and water pollution, waste generation and pressures on biodiversity and wildlife habitats. Refining mined ores into metal is energy and water intensive. Fossil fuel exploitation result in pollution and habitat disruption at the extraction sites and carbon dioxide emissions when used in combustion. Non-energy uses of fossil fuels e.g. in plastics or chemicals have a different set of environmental implications, including the pollution of environmental systems by persistent plastic waste (Box 8.1) and toxic contamination by chemicals. Unsustainable production of biomass, i.e. farming, fishing and forestry, can contribute to land cover changes, the loss of ecosystem services, biodiversity loss and soil degradation. Unsustainable deforestation can lead to increased soil erosion, habitat destruction and loss of biodiversity. It also depletes carbon sinks and thus contributes to global warming.

Table 8.1. Potential environmental impacts by material group

Material Group	Potential environmental impacts
Biomass (for food and feed)	Intensification of land use, land cover change, soil degradation, groundwater contamination, disintegration of nutrient cycles, food chain contamination through pesticides, acidification, loss of biodiversity, habitat loss, water use, eutrophication
Wood	Intensification of land use, soil erosion, loss of biodiversity, forest degradation, habitat alteration, carbon sink depletion, desertification, alteration of watersheds
Fossil energy carriers	Air pollution, carbon dioxide emissions, habitat alteration, overburden, toxic chemicals for processing, water usage
Metals and metal ores	Irreversible ecosystem change (entropy generation), toxicity, habitat alteration, mining overburden, air emissions, water usage, tailings, radioactivity
Industrial minerals	Irreversible ecosystem change (entropy generation), toxicity, habitat alteration, mining overburden, air emissions, waste water, tailings
Construction minerals	Loss of biodiversity, habitat alteration, soil compaction, CO_2 emissions (e.g. cement manufacturing), transport intensity, sealing of land area, soil compaction

Source: OECD (2015[2]).

Additionally, the environmental consequences of resource use vary throughout the materials lifecycle. The extraction of raw materials generally requires energy and water, generating pollution and waste, and often permanently or temporarily alters the surrounding habitat. The processing and consumption stages generate pollution and waste, both processing-related and accidental. There is also potential for negative environmental impacts during transportation, including accidents or leakages, such as oil spills. Transport fuel consumption generates atmospheric emissions, which can contribute to air pollution and climate change.

Products can also have various environmental impacts at the end of their life, if waste is not managed adequately. Additionally, environmental impacts can be widely distributed geographically if resources are traded internationally, if resource reservoirs extend beyond borders or if impacts flow across countries (e.g. water or air pollutants).

There are also interesting trade-offs between resources. One example is the potential to use different construction materials, each with their own environmental impacts. Another is the trade-off between land use change and the use of fertilisers (such as potassium and phosphate) to meet a growing demand for food. Such trade-offs imply that policy responses should look at the full range of environmental consequences and avoid a silo approach where one environmental problem is substituted by another.

A full quantitative assessment of the environmental consequences of materials use at the global level is not feasible, due to the lack of data as well as the large amount of environmental impacts involved when considering all materials and resources. This section therefore focuses on quantifying impacts for climate change and for the broad environmental impacts of selected metals, concrete and sand and gravel. In addition, Box 8.1 highlights the issue of plastics, which – as a processed material – is not explicitly considered in this report.

Box 8.1. The rise and rise of plastics

While this report does not assess plastics in any detail, they represent an important family of materials that have become ubiquitous in everyday life. Plastics are key inputs in a wide range of applications, from the films used in product and food packaging and the fibres used in textiles, to the durable plastics used in vehicles and construction. Global plastics production reached 407 Mt in 2015, making plastics more widespread than the production of paper (400 Mt), fish (200 Mt), or aluminium (57 Mt) (WWF, 2018[3]; World Bank, 2018[4]; USGS, 2016[5]). Looking ahead, it has been suggested that the production and use of plastics could quadruple by 2050 (EMF, 2017[6]).

Plastics have gathered increasing attention recently due to their associated environmental impacts. Plastics production and disposal are responsible for significant GHG emissions and, when poorly managed, pollute the natural environment, including in the oceans. Plastic pollution is present in all the world's major ocean basins, and it is estimated that an additional 5 to 13 Mt are introduced every year (Jambeck et al., 2015[7]). The ecosystem damages and risks to human health are only beginning to emerge, but are of clear concern given the longevity of plastics and the highly visible impact on marine life. The economic costs of plastic pollution in the marine environment, including those associated with diminished fisheries, reduced tourism, and time spent cleaning up beaches, have been conservatively estimated at USD 13 billion per year (UNEP, 2014[8]).

The environmental impacts of plastics can be addressed in a number of ways. These include the promotion of waste prevention strategies (e.g., the introduction of reusable plastic products), the substitution of alternative, less environmentally harmful materials (e.g., certain types of biodegradable plastics), the development of more effective waste collection and treatment systems, and the design of plastics that are relatively easily recycled. Recent OECD work has focused particularly on the role that better functioning markets for recycled plastics could play in stimulating higher plastic collection and recycling rates (OECD, 2018[9]). This would help to reduce the diffusion of plastics pollution in the environment while continuing to allow the beneficial aspects of plastics to be realised.

8.2. Reducing greenhouse gas emissions is strongly linked to materials use policies

In the central baseline scenario, which assumes no new policies are implemented, global GHG emissions from all sources are projected to increase over time (from 41 Gt CO_2 equivalent in 2011 to 75 Gt CO_2 equivalent in 2060), as shown in Figure 8.2.[1] In the absence of further policy action, these emissions will lead to climate impacts with substantial economic consequences (see Annex 8.A).

Emissions are projected to remain stable within the OECD region, as shown in Figure 8.2. But the catch-up of emerging and developing economies is projected to increase emissions, with the biggest increase – but not the biggest absolute contribution – foreseen for Africa (grouped with the Middle East in the figure).[2]

Combustion of fossil fuels remains the main driver of CO_2 and total GHG emissions, although emissions of all gases are projected to increase. But with the projected widespread energy efficiency improvements, combustion emissions especially are projected to grow less rapidly than economic activity. Similarly, agricultural emissions are projected to grow less rapidly than GDP, reflecting the diminishing share of agriculture in the global economy. However, emissions per unit of agricultural production are hardly declining in this baseline projection.

Figure 8.2. Global GHG emissions are projected to grow in the central baseline scenario

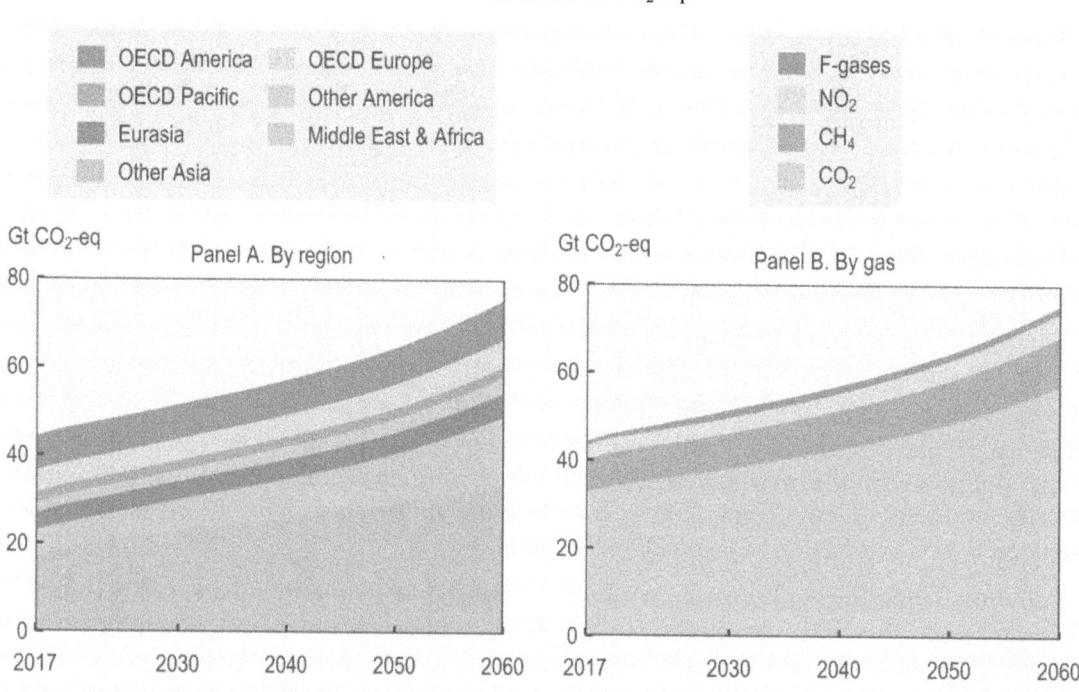

Note: The figures exclude emissions from land use, land use change and forestry (LULUCF).
Source: OECD ENV-Linkages model.

StatLink https://doi.org/10.1787/888933885581

The central baseline scenario does not include the policies that would be necessary to meet the ambitions of the Paris Agreement, including the Nationally Determined

Contributions (NDC) and the "well below two degrees" objectives. This does not indicate that these objectives are unattainable, but rather that additional policy efforts are required to meet them. For example, IEA (2017[10]) project that the NDCs are met in their New Policies Scenario. This scenario projects a reduction in coal use of around 20% and global CO_2 emissions that are 6 Gt lower in 2040 compared to the Current Policies Scenario. The latter (CPS) scenario is used to calibrate the central baseline scenario (see Box 2.4 in Chapter 2).³

Most emissions are directly or indirectly linked to the use of materials, and, more specifically, to what OECD (2012[1]) refers to as "materials management" (Figure 8.3). Materials management emissions come mostly from the combustion of fossil fuels for energy supply, but also include emissions from agriculture, emissions that are linked to the production of manufacturing goods, and emissions from construction. Emissions related to materials management are projected to increase from 30 Gt in 2017 to 49 Gt CO_2-eq in 2060. Emissions not related to materials management include those from transport, as well as from end users (households, services, and private transportation). These are a smaller but still relevant share of overall emissions. Non-materials management emissions are projected to represent an increasing share of overall emissions by 2060 – from 31% of overall emissions in 2011 to 35% in 2060 – reflecting servitisation and other sectoral changes, as well as energy efficiency improvements, especially in non-OECD countries.

Figure 8.3. Greenhouse gas emissions are projected to increase

Global GHG emissions in Gt CO_2-eq.

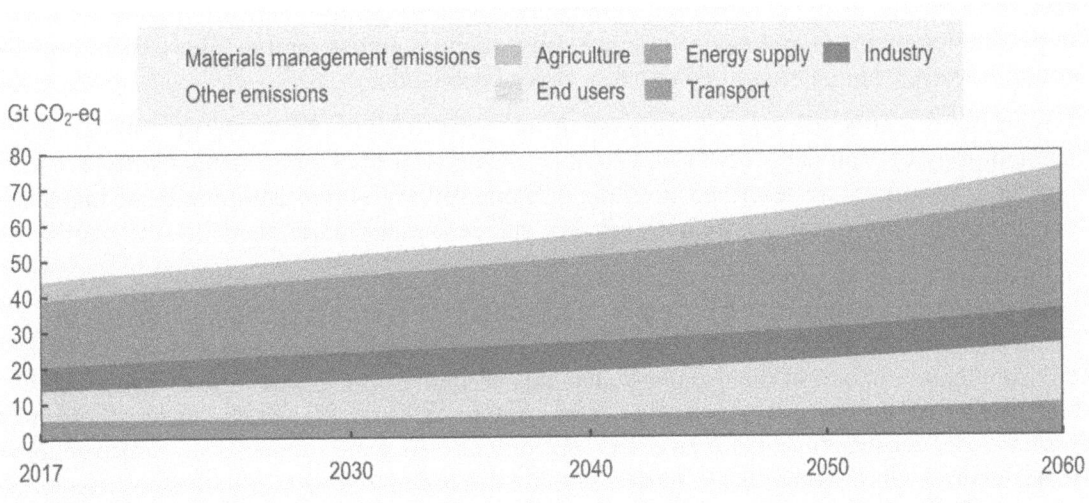

Note: Emissions are differentiated between those linked to materials management and those that are not, based on previous OECD work on materials management within OECD countries (OECD, 2012[1]). For example, fossil fuel combustion by households is not attributed to materials management, while all industrial activity is. End users includes services and households as well as private transportation (light duty vehicles).
Source: OECD ENV-Linkages model.

StatLink ⟶ https://doi.org/10.1787/888933885600

Reducing GHG emissions is strongly linked to policies to manage materials use; for example reducing fossil fuel use directly contributes to lowering GHG emissions and reducing materials use. However, boosting the production of renewable energy as substitute for fossil fuels in order to reduce GHG emissions can increase the use of other materials, as renewable technologies rely more heavily on metals. Similarly, emissions from transport can be reduced with the greater use of electric cars, but that will also lead to a higher demand for specific materials. Finally, there are trade-offs in designing vehicles between optimising for energy efficiency or for recyclability. But perhaps most importantly, synergies between resource efficiency and circular economy policies on the one hand and climate change mitigation policies on the other can be exploited as both sets of policies aim to shift the economy away from resource-intense activities towards "cleaner" production methods and commodities.

8.3. The increase in materials use will exacerbate environmental impacts

8.3.1. The environmental impacts of extraction and processing materials are diverse

Environmental impacts are associated with different parts of the life cycle of resource use: from extraction to processing to discarding as waste. Life cycle analysis (LCA) can be used to assess the direct and indirect environmental consequences of resource use along their life cycle stages.

This report presents an assessment of the environmental impacts of global extraction and processing (in short, production) for two types of materials between 2010 and 2060: 7 metals (iron, aluminium, copper, zinc, lead, nickel and manganese) as well as the construction material concrete.[4] This section focuses on the metals, while the next focuses on the construction materials. One important caveat to be kept in mind throughout the section is that the analysis is based on global averages. In reality, there are big differences in production methods and environmental consequences across regions; these differences could, however, not be quantified due to a lack of robust data.

To perform a quantification of the environmental impacts, this report follows the methodology of Van der Voet et al. (2018[11]).[5] The environmental impacts studied here include 9 indicators as described in Table 8.2, quantifying global environmental impacts per kilogramme of metal use for both primary and secondary materials.[6]

To avoid accounting problems, the indicators refer to cradle-to-gate impacts. Cradle-to-gate impacts cover the upstream portion of the life cycle (extraction and processing). As a result, the impacts associated with the use and end-of-life phases are not considered.[7] The environmental impacts at the use phase depends crucially on the type of use of the metals and can be either positive or negative. For instance, the use of nickel can lengthen the lifetime of construction materials and so reduce lifecycle impacts of construction materials use. On the other hand, metals that are discarded and end up in the environment after single use may contribute further to pollution.

Additional assumptions are made in the analysis for ore grade decline (for copper, lead, zinc, nickel) and decrease in the energy intensity of the production chain (for aluminium, iron and steel, copper) over the period 2015-2060. Furthermore, ongoing efficiency improvements in the extraction and processing sectors may not reduce the per-kg impacts, but reduce the impact per unit of output of the processed material (e.g. a dollar of steel).

CHAPTER 8. PROJECTIONS OF THE ENVIRONMENTAL CONSEQUENCES OF MATERIALS USE | 189

Table 8.2. Environmental impact categories and indicators

Environmental impact	Nature of impact	Indicator (unit)
Acidification	Corrosive impact that pollutants such as sulphur dioxide (SO_2) and Nitrous Oxides (NO_x) have on soil, groundwater, surface waters, biological organisms, ecosystems and materials (buildings).	Emissions of acidifying gases to the air (kg SO_2 eq)
Climate change	Anthropogenic emissions causing the temperature of the Earth surface to rise and leading to several impacts on the environment (e.g. sea level rise, extreme weather events) and the economy (e.g. agriculture and ecosystem services).	GHG emissions to the air (kg CO_2-eq)
Cumulative energy demand	Energy footprint (total energy use along the production chain of a material).	Energy use (MJ)
Eutrophication	Potential impacts of excessively high levels of macronutrients (such as nitrogen (N) and phosphorus (P)). Nutrient enrichment may cause undesirable shift in species composition and elevated biomass production in ecosystems and affects sources suitable for drinking water.	Emissions of nutrients to air, water and soil (kg PO_4-eq)
Freshwater aquatic ecotoxicity	Impacts of toxic substances on species in freshwater aquatic ecosystems.	Emissions of toxic substances to air, water and soil (kg 1,4-dichlorobenzene eq)
Human toxicity	Impacts on human health of toxic substances in the environment, either by inhalation or via the food chain. Such impacts cover widely varying symptoms reaching from irritation to mortality.	Emissions of toxic substances to air, water and soil (kg 1,4-dichlorobenzene eq)
Land use	Land surface used to produce the resource (e.g., area occupied by a mine).	Land use (m^2)
Photochemical oxidation	Formation of reactive chemical compounds such as ozone by the action of sunlight on certain primary air pollutants, sometimes visible as smog.	Emissions of substances (VOC, CO) to air (kg ethylene eq)
Terrestrial ecotoxicity	Impacts of toxic substances on species in terrestrial ecosystems.	Emissions of toxic substances to air, water and soil (kg 1,4-dichlorobenzene eq)

8.3.1. The production of primary materials is more polluting than that of secondary materials

According to this analysis, primary copper and nickel production are the ones with the highest impacts per kilogramme of produced metals, for the selected environmental impacts. The per kg environmental impact values for 2015 are summarised in Figure 8.4. They are expressed as an index: for each environmental indicator, the metal whose production has the largest impact gets a value of 1.

It is possible that extraction and processing of specific metals not investigated in this report (e.g. rare earth elements) are more polluting than the metals presented here. Due to a lack of robust data, the global environmental consequences of production of other metals cannot be assessed. Hence, the term "most polluting" should be interpreted with care, as it is only in relative terms, i.e. in comparison to the other investigated materials.

Primary nickel production has the highest per kg values for 5 of the 10 indicators (acidification, climate change, cumulative energy demand, photochemical oxidation, terrestrial ecotoxicity), and also high values for land use. A driving factor for these impacts is that its production requires a large amount of energy, with consequences for e.g. GHG emissions.

Figure 8.4. Per kg environmental impacts are higher for primary than for secondary materials

Normalised index value (highest impact normalised to 1) of different environmental impacts for 2015

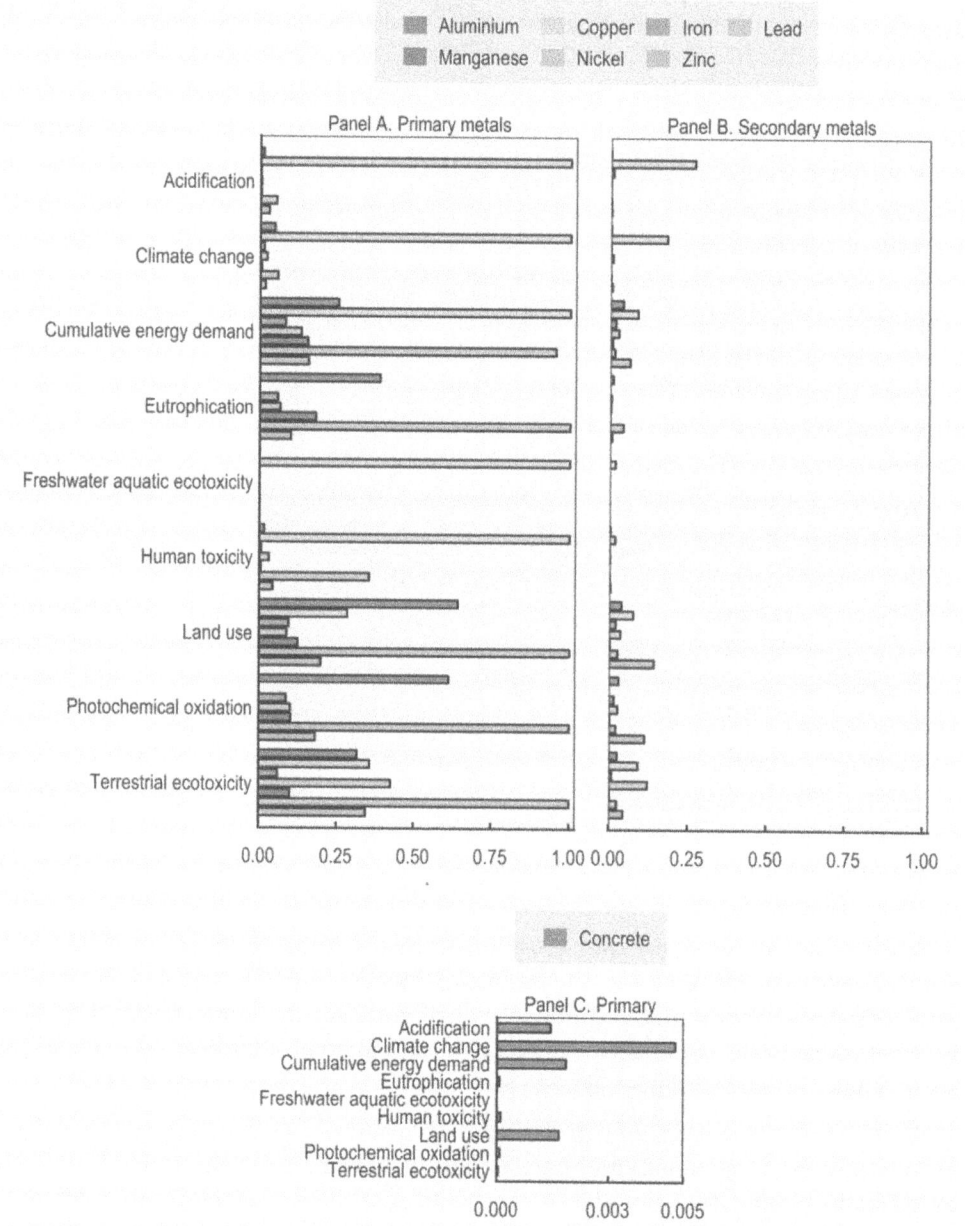

Source: CML's CMLCA model, based on the central baseline scenario of the OECD ENV-Linkages model.

StatLink https://doi.org/10.1787/888933885619

Primary copper production has the highest per kg impact for the other 4 impacts (eutrophication, freshwater aquatic ecotoxicity, human toxicity and land use). It is the only metal in this list whose production has significant impacts on freshwater aquatic ecotoxicity. For eutrophication and freshwater aquatic ecotoxicity, primary nickel production is also relatively polluting, while the per kg environmental impacts of production of other metals are much less.

Aluminium production, while generally not as polluting as copper or nickel production, shows high impacts stemming from its extraction and processing on many indicators. Especially its impact on photochemical oxidation is close to that of nickel. Its impact on climate change and cumulative energy demand is also higher than half that of nickel (which is the highest in terms of per kg impacts).

For some environmental impacts, some metals are much less polluting than others. For example, the water pollution related impacts (freshwater ecotoxicity and eutrophication) are much more significant for production of nickel and copper than for the production of other metals. The impacts directly or indirectly related to energy use are, however, more evenly spread and significant for all metals.

The environmental impact assessment also assesses the per kg environmental impacts associated with secondary materials production (panel B in Figure 8.4). This figure shows that the impacts tend to be an order of magnitude lower than those for primary-based production. The impacts of secondary copper production are smaller by a factor 4-60 when compared to primary, and the impacts of secondary nickel production are 25-300 times smaller than primary.

Nonetheless, some of the impacts from secondary production are not negligible compared to primary. Secondary zinc production has relatively high impacts on energy demand and thus climate change: the values for secondary are more than half those of primary. The photochemical oxidation impacts of secondary production for lead and zinc are also relatively high, amounting to roughly one-third of the value for primary production. In one case secondary production is even more polluting than primary: the terrestrial ecotoxicity impact of secondary iron is almost 5 times higher that of primary iron, albeit still much lower than that of nickel (as shown in Figure 8.5).

The per kg impacts of concrete production are much lower than those of metals production. Compared to the metal production with the highest impacts, the environmental impacts per kg of material are two to three orders of magnitude smaller (i.e. less than 1% of the most polluting metal production).

8.3.2. The per kg environmental impacts evolve over time

Overall, the per kg environmental impacts are projected to decrease over time. Figure 8.6 shows that this is the case for many materials. These improvements stem from efficiency improvements in extraction and processing, the gradual adoption of less polluting production processes as well as the evolution of the energy mix that is used; a significant part of this is the transition towards more renewables in elect electricity production, which lowers e.g. greenhouse gas emissions related to electricity use in materials extraction and processing.

For some metals, these improvements are countered by the increased environmental consequences caused by the projected declining in ore grades (Van der Voet et al., 2018[11]). This holds for the primary processing of copper, lead, nickel and zinc. As ore grades decline, more energy is needed at the extraction phase, leading to larger environmental impacts.

The per kg impacts of secondary materials production improve for all the presented metals, under the assumption that the recycling of scrap materials does not become harder over time. The decrease in the environmental impacts of secondary production follows technological improvements and the evolution of the energy mix.

Figure 8.5. Secondary materials lead to much lower environmental impacts

Normalised index value (highest impact normalised to 1) of different environmental impacts for selected metals for 2015

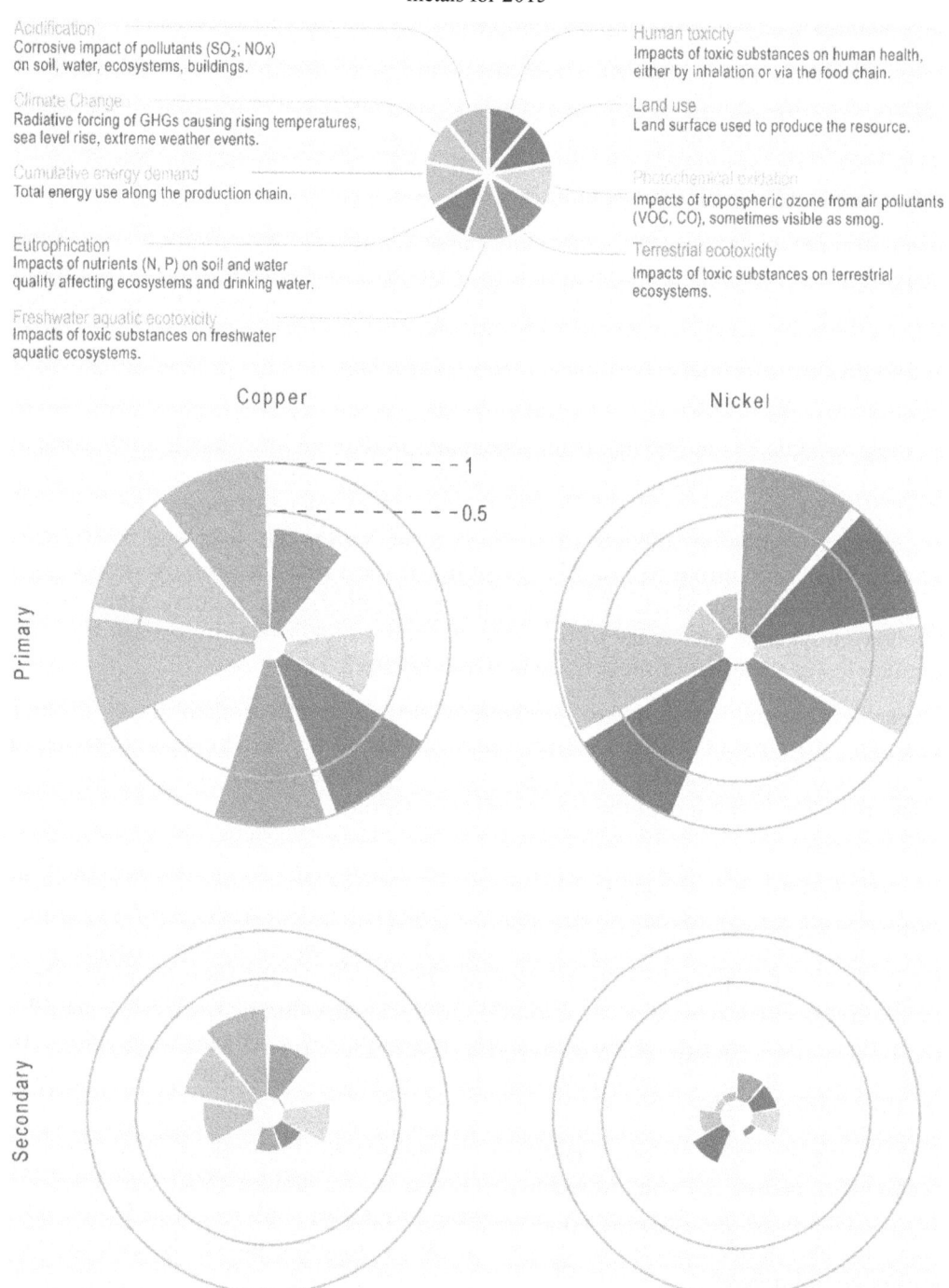

Source: CML's CMLCA model, based on the central baseline scenario of the OECD ENV-Linkages model.

StatLink https://doi.org/10.1787/888933885638

CHAPTER 8. PROJECTIONS OF THE ENVIRONMENTAL CONSEQUENCES OF MATERIALS USE | 193

Figure 8.6. Per kg environmental impacts decrease over time, except when ore grades degrade significantly

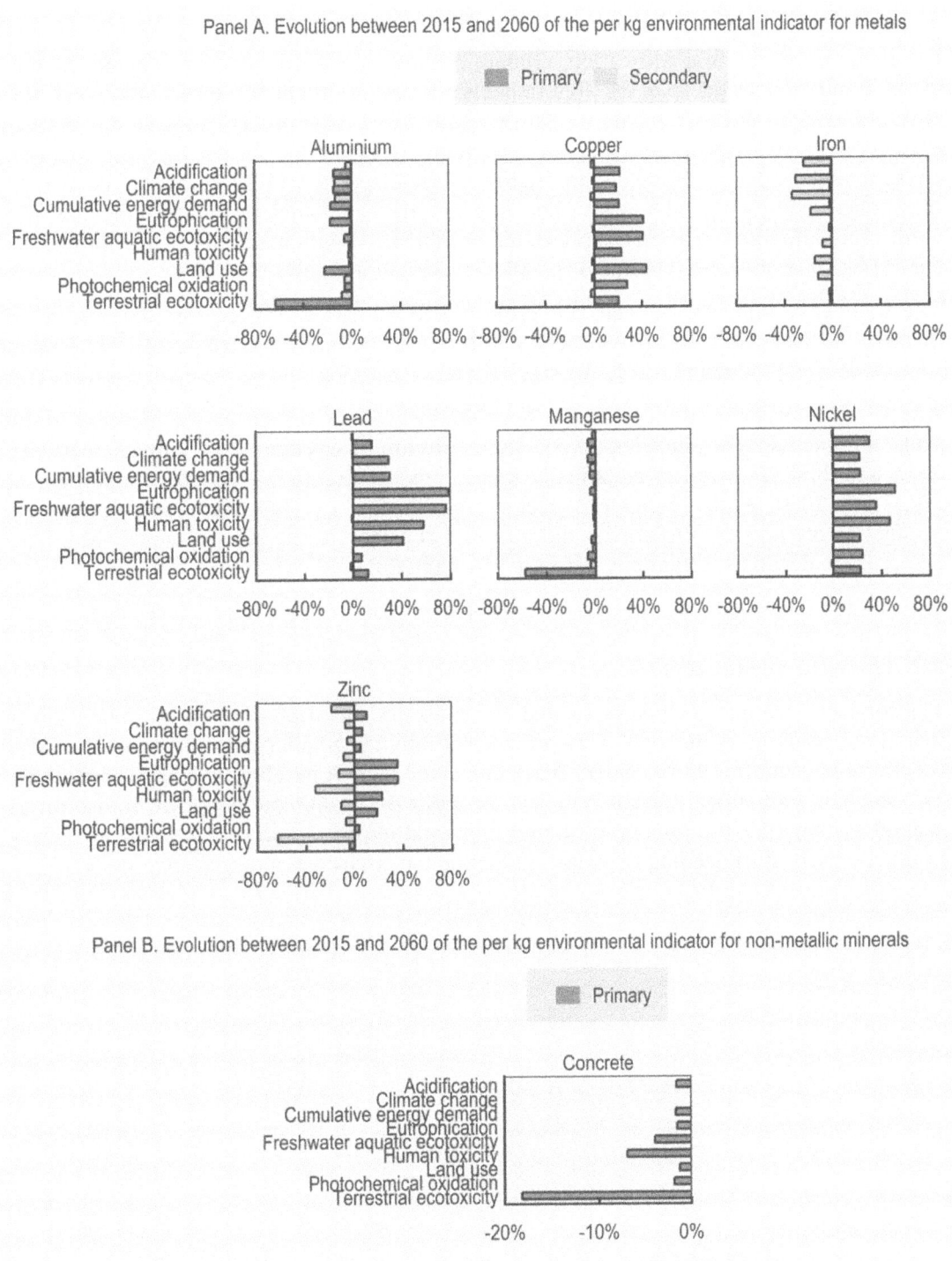

Source: CML's CMLCA model, based on the central baseline scenario of the OECD ENV-Linkages model.

StatLink https://doi.org/10.1787/888933885657

8.3.3. The environmental consequences of metals are wide-ranging

The scale of iron ore extraction and processing implies that it is projected to have the largest total impact on acidification, climate change, cumulative energy demand, human toxicity and land use, despite the high impacts per unit of output of copper and nickel. For eutrophication, photochemical oxidation and both ecotoxicity indicators, copper extraction and processing are especially problematic. Figure 8.7 shows the projected total environmental impacts of metals production, normalized to 1 in 2060 for the metal with the highest environmental impact. The impacts are calculated by multiplying the unit values (per kilogramme of metal) presented in Figure 8.4 with materials projections of the central baseline scenario (volume of metal use in kilogrammes).

The total environmental impacts are projected to increase strongly for the considered metals, and increase by 250% to 400% with respect to their current levels.[8] The main polluter is iron production due to the size of total production.[9]

Almost all impacts are projected to at least double, mostly driven by the increase in the scale of materials use (see Chapter 5). The evolution of environmental impacts between 2015 and 2060 by metal is graphically represented in Figure 8.7 via the comparison of 2015 and 2060 levels. Terrestrial ecotoxicity impacts of aluminium and manganese production are projected to remain more or less constant, as the per kg impact values are projected to decline substantially between 2015 and 2060.

The largest increases in impacts are projected for copper, zinc, lead and nickel production. Especially large increases are projected for eutrophication, human toxicity and freshwater aquatic ecotoxicity.

The total environmental impacts related to secondary metal production are much lower than those for primary metals.[10] This reflects not only the smaller per-unit values, but also the limited penetration of secondary materials in the central baseline scenario (cf. Chapter 5), which is built upon current policies. As policies further ramp up the transition to secondary materials use, this ratio of environmental impacts from primary and secondary production will shift, leading to overall reductions in environmental impacts.

Figure 8.7. Environmental impacts of selected metals are in most cases projected to more than double by 2060

Total environmental impacts in 2015 (lighter shaded area) and 2060 (full coloured area), index 1 for most polluting material in 2060

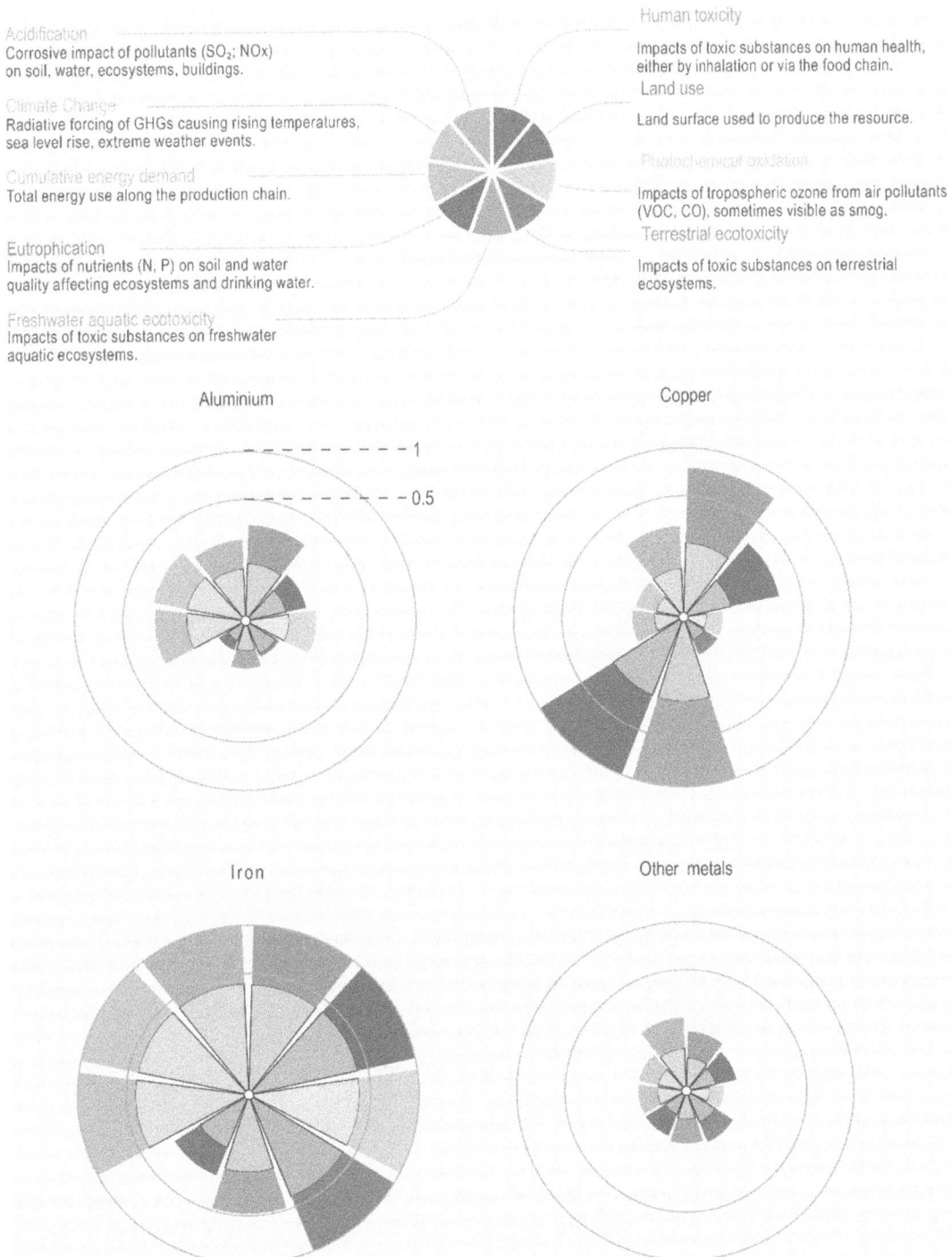

Figure 8.7. Environmental impacts of selected metals are in most cases projected to more than double by 2060 (*continued*)

Panel B. Total environmental impacts for selected metals primary vs. secondary

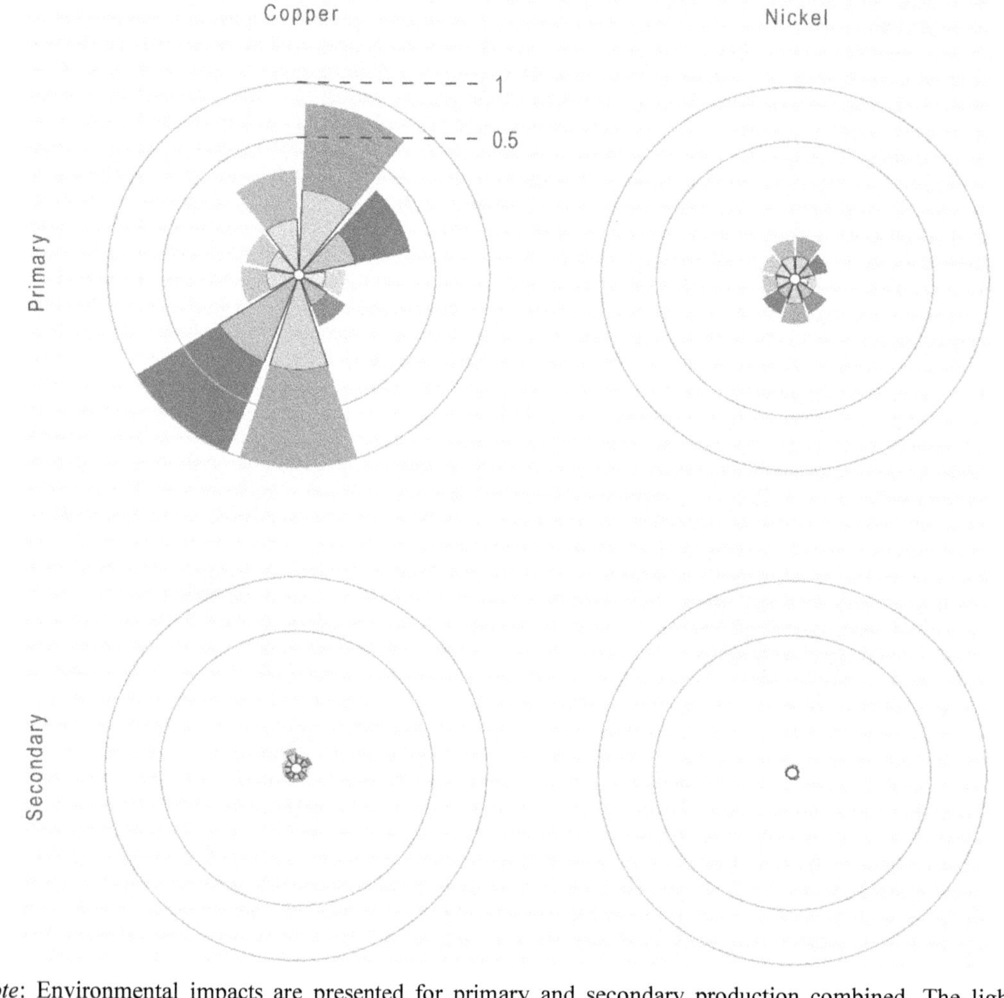

Note: Environmental impacts are presented for primary and secondary production combined. The lighter shading represents the value in 2015; the full coloured area reflects values in 2060. Impacts for "Other metals" reflect the combined impacts of lead, manganese, nickel and zinc.
Source: CML's CMLCA model, based on the central baseline scenario of the OECD ENV-Linkages model.

StatLink https://doi.org/10.1787/888933885676

8.3.4. The large-scale use of construction materials brings significant environmental consequences

While the per kg impacts of construction materials are lower than those of metals, the quantities of these materials used are so high that the overall impact they have on the environment cannot be ignored. The projected environmental impacts of concrete show that they have a large effect on most indicators (see Figure 8.7). Given the large increase in the use of construction materials in the baseline projection, the environmental impacts of construction materials should be further investigated.

CHAPTER 8. PROJECTIONS OF THE ENVIRONMENTAL CONSEQUENCES OF MATERIALS USE | 197

The environmental impacts of concrete production further increase to 2060, not least for acidification, climate change, cumulative energy demand and land use. In particular, GHG emissions related to concrete production double to 2060, and are as large as the emissions of the 7 selected metals taken together.

Figure 8.8. Projected environmental impacts of selected materials in 2060

Total environmental impacts in 2015 (lighter shaded area) and 2060 (full coloured area), index 1 for most polluting material in 2060

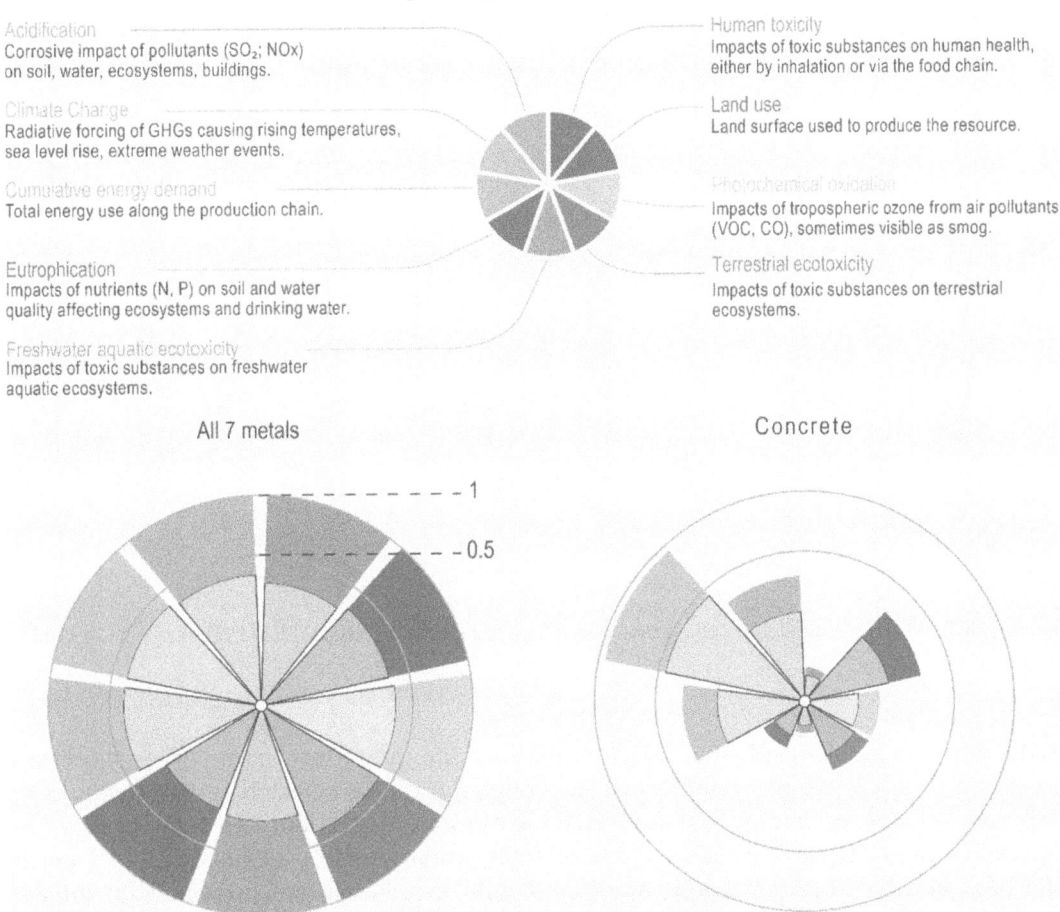

Note: Environmental impacts are presented for primary and secondary production combined. The lighter shading represents the value in 2015; the full coloured area reflects values in 2060.
Source: CML's CMLCA model, based on the central baseline scenario of the OECD ENV-Linkages model.

StatLink https://doi.org/10.1787/888933885695

8.3.5. Materials extraction and production forms a significant share of total environmental impacts

Comparing the importance of the different environmental impacts is not easy. Comparing the contribution of the impacts of the production of these materials to the overall environmental impact of all economic activity together (for instance the contribution of concrete production to global GHG emissions) helps understand the importance of materials management for limiting environmental problems. To that end, Figure 8.8 presents the share of the problem for selected materials and indicators.

Figure 8.9. Metals and concrete represent a significant share of the environmental impacts

Shares of selected global environmental impacts

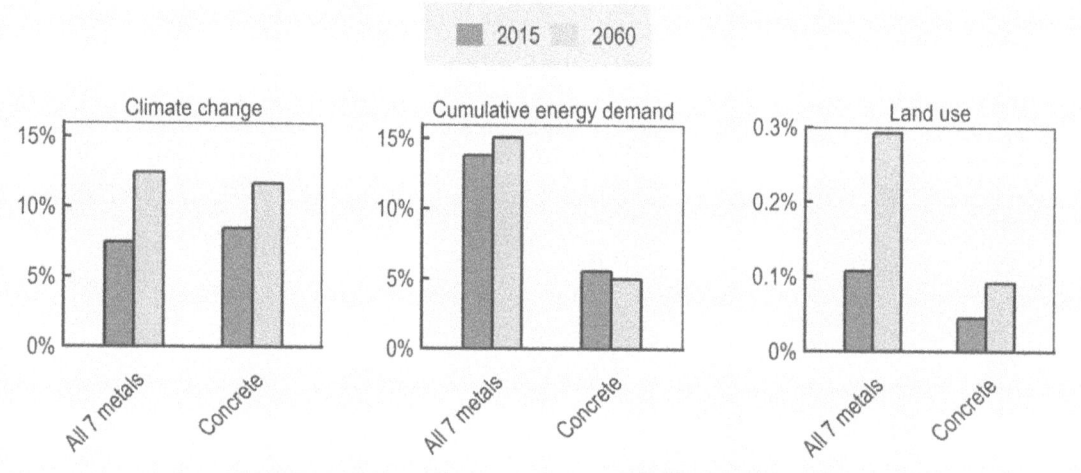

Source: CML's CMLCA model, based on the central baseline scenario of the OECD ENV-Linkages model.

StatLink https://doi.org/10.1787/888933884403

Reducing energy demand and GHG emissions are important parts of a successful transition to a low carbon economy. The production of the selected seven metals is an important source of energy use and GHG emissions. By 2060, the total energy use corresponding to these 7 metals is 15% of total primary energy demand (134 EJ out of 887 EJ), and the GHG emissions associated with metals use are projected in 2060 to amount to 12% of total projected GHG emissions. This corresponds to about one third of total industrial emissions globally. The climate change impact of concrete is projected to amount to 12% of total GHG emissions in 2060. In addition, concrete production represents about 5% of cumulative energy demand.

But the environmental impacts of production of these selected materials is not always a major part of the total environmental challenge. For example, the projected land use impact of these metals corresponds to 0.3% of total land.[11] The growth between 2015 and 2060 corresponds to almost a tripling of the land use footprint for these metals extraction and processing (from 139 to 379 thousand km^2).

The analysis presented here is tentative, and hides large differences in environmental consequences across regions. Nonetheless, it highlights that materials management can play a significant role in environmental policies. As identified in this section, an important strategy to reduce the environmental impacts of metal extraction and processing is to transition towards secondary supply.

Other solutions include the development of innovative production processes (see Smil (2016[12]) for an example for iron). The reduction of the use of metals, and especially iron and steel, could also help, but may be difficult to achieve given the global development trends presented in Chapters 3 and 4. The substitution of other materials for metals might be a way forward, but careful consideration has to be given to the trade-offs in environmental impacts (using wood may incur an additional burden on deforestation for instance).

Notes

[1] The emission projections in the central baseline scenario are consistent with the projections in recent literature as summarised by the IPCC's Fifth Assessment Report (IPCC, 2014[13]) and (Riahi et al., 2015[14]), albeit at the lower end, as recent developments in renewables and economic growth have led to a downward adjustment of emissions projections (see Figure 8.A.1).

[2] These differences between OECD and non-OECD countries depend on the fact that emissions are attributed to the relevant production process. Emissions related to production of commodities in a non-OECD country that are then exported for consumption in OECD countries are not attributed to the OECD region; these flows are quite large.

[3] Some of these policy developments have already started. In line with IEA (2017[10]), policies introduced after mid 2017 have not been included in the baseline scenarios.

[4] The estimation of the impacts of minerals on health and the environment are the topic of discussion and research. Regardless of the results presented in this report, more research is ongoing to improve environmental impact estimations for inorganic materials in general, and metals in particular, that could improve the projections given in this study in the future.

[5] More information on the methodology and the indicators used in this analysis can be found in Guinee (2002[17]).

[6] See Annex 8.A for more details.

[7] The cradle-to-gate impacts are calculated using the LCA software tool CMLCA (http://www.cmlca.eu/). The calculations are based on process descriptions compiled by Ecoinvent (https://www.ecoinvent.org/database/database.html, database version 2.2). The use of dispersion and fate models for determining the toxicity of metals is under discussion and better assessment tools are in development; the results presented here are thus a snapshot of the available information at the time of the analysis. It should however be noted that the largest toxic effects are not directly related to the loss of metal to the environment, but to the use of chemicals and fossil fuels in extraction and processing.

[8] These projections of environmental impacts differ from Van der Voet et al. (2018[11]) not only in scope (more indicators, longer time horizon), but also in the projected evolution of the per kg impacts as a result of different assumptions used in the calculations. Most importantly, the projections of total metals use between 2015 and 2060 are not the same, as the central baseline scenario used here is based on more recent forecasts and detailed economic projections (see Chapter 2). Thus, the numbers differ; the qualitative conclusions are however largely the same.

[9] For iron, a change in the energy mix would not solve the environmental problem, as most of the environmental impacts derive from the coal used in the process. For other metals, changing the energy mix would significantly reduce their environmental impacts.

[10] Table 8.A.1 in Annex 8.A provides detailed results for primary and secondary materials.

[11] To give an equivalent, it corresponds to the total surface of New Zealand, Burkina Faso or the Philippines.

References

Derwent, R., M. Jenkin and S. Saunders (1996), "Photochemical ozone creation potentials for a large number of reactive hydrocarbons under European conditions", *Atmospheric Environment*, Vol. 30/2, pp. 181-199, http://dx.doi.org/10.1016/1352-2310(95)00303-G. [23]

Derwent, R. et al. (1998), "Photochemical ozone creation potentials for organic compounds in northwest Europe calculated with a master chemical mechanism", *Atmospheric Environment*, Vol. 32/14-15, pp. 2429-2441, http://dx.doi.org/10.1016/S1352-2310(98)00053-3. [22]

EMF (2017), *Rethinking the future of plastics and catalysing action*, https://www.ellenmacarthurfoundation.org/assets/downloads/publications/NPEC-Hybrid_English_22-11-17_Digital.pdf. [6]

Guinée, J. (2002), *Handbook on life cycle assessment : operational guide to the ISO standards*, Kluwer Academic Publishers. [17]

Heijungs, R. et al. (1992), *Environmental life cycle assessment of products: guide and backgrounds*, Centre of Environmental Science (CML), Leiden University, Leiden, The Netherlands, https://openaccess.leidenuniv.nl/handle/1887/8061 (accessed on 20 September 2018). [20]

Huijbregts, M. (2000), *Priority Assessment of Toxic Substances in the frame of LCA. Time horizon dependency of toxicity potentials calculated with the multi-media fate, exposure and effects model USES-LCA.*, Institute for Biodiversity and Ecosystem Dynamics, University of Amsterdam, http://www.leidenuniv.nl/interfac/cml/lca2/. [24]

Huijbregts, M. (1999), *Life-cycle impact assessment of acidifying and eutrophying air pollutants: Calculation of equivalency factors with RAINS-LCA*, Interfaculty Department of Environmental Science, Faculty of Environmental Science, University of Amsterdam, The Netherlands, https://media.leidenuniv.nl/legacy/Life-cycle%20impact%20assessment.pdf (accessed on 20 September 2018). [19]

Huijbregts, M. (1999), *Priority assessment of toxic substances in LCA. Development and application of the multi-media fate, exposure and effect model USES-LCA*, IVAM environmental research, University of Amsterdam. [25]

IEA (2017), *World Energy Outlook 2017*, OECD Publishing, Paris/IEA, Paris, http://dx.doi.org/10.1787/weo-2017-en. [10]

IPCC (2014), *Climate Change 2014: Mitigation of Climate Change. Contribution of Working Group III to the Fifth Assessment Report of the Intergovernmental Panel on Climate Change*, Cambridge University Press, Cambridge, United Kingdom and New York, NY, USA, https://www.ipcc.ch/report/ar5/wg3/. [13]

Myhre, G., D. Shindell, F.-M. Bréon, W. Collins, J. Fuglestvedt, J. Huang, D. Koch, J.-F. Lamarque, D. Lee, B. Mendoza, T. Nakajima, A. Robock, G. Stephens, T. (ed.) (2013), *Anthropogenic and Natural Radiative Forcing. In: Climate Change 2013: The Physical Science Basis. Contribution of Working Group I to the Fifth Assessment Report of the Intergovernmental Panel on Climate Change*, Cambridge University Press, Cambridge, United Kingdom and New York, NY, USA.. [18]

Jambeck, J. et al. (2015), "Marine pollution. Plastic waste inputs from land into the ocean.", *Science (New York, N.Y.)*, Vol. 347/6223, pp. 768-71, http://dx.doi.org/10.1126/science.1260352. [7]

Jenkin, M. and G. Hayman (1999), "Photochemical ozone creation potentials for oxygenated volatile organic compounds: sensitivity to variations in kinetic and mechanistic parameters", *Atmospheric Environment*, Vol. 33/8, pp. 1275-1293, http://dx.doi.org/10.1016/S1352-2310(98)00261-1. [21]

OECD (2018), *Improving Markets for Recycled Plastics - Trends, Prospects and Policy Responses*, Organization for Economic Cooperation and Development. [9]

OECD (2015), *Material Resources, Productivity and the Environment*, OECD Green Growth Studies, OECD Publishing, Paris, http://dx.doi.org/10.1787/9789264190504-en. [2]

OECD (2015), *The Economic Consequences of Climate Change*, OECD Publishing, Paris, http://dx.doi.org/10.1787/9789264235410-en. [16]

OECD (2012), *Greenhouse gas emissions and the potential for mitigation from materials management within OECD countries*, OECD, https://one.oecd.org/document/ENV/EPOC/WGWPR(2010)1/FINAL/en/pdf (accessed on 22 May 2018). [1]

OECD (2012), *OECD Environmental Outlook to 2050: The Consequences of Inaction*, OECD Publishing, Paris, http://dx.doi.org/10.1787/9789264122246-en. [15]

Riahi, K. et al. (2015), "The Shared Socioeconomic Pathways and their energy, land use, and greenhouse gas emissions implications: An overview", *Global Environmental Change*, http://dx.doi.org/10.1016/j.gloenvcha.2016.05.009. [14]

Smil, V. (2016), *Still the Iron Age: Iron and Steel in the Modern World*, http://dx.doi.org/10.1016/C2014-0-04576-5. [12]

UNEP (2014), "Valuing Plastics: The Business Case for Measuring, Managing and Disclosing Plastic Use in the Consumer Goods Industry", http://www.gpa.unep.org (accessed on 24 July 2018). [8]

USGS (2016), *Aluminum Legislation and Government Programs*, https://minerals.usgs.gov/minerals/pubs/commodity/aluminum/myb1-2015-alumi.pdf (accessed on 28 March 2018). [5]

Van der Voet, E. et al. (2018), "Environmental Implications of Future Demand Scenarios for Metals: Methodology and Application to the Case of Seven Major Metals", *Journal of Industrial Ecology*, http://dx.doi.org/10.1111/jiec.12722. [11]

World Bank (2018), *Total fisheries production (metric tons) | Data*, https://data.worldbank.org/indicator/ER.FSH.PROD.MT (accessed on 28 March 2018). [4]

WWF (2018), *Pulp and paper | WWF*, http://wwf.panda.org/about_our_earth/deforestation/forest_sector_transformation/pulp_and_paper/ (accessed on 28 March 2018). [3]

Annex 8.A. Detailed results and supplementary materials

8.A.1. Comparing the GHG projections to the literature

The global GHG emission projections can be compared to earlier projections, especially the range of baseline projections used for the Fifth Assessment Report of the IPCC (IPCC, 2014[13]), the projections of the Shared Socioeconomic Pathways (Riahi et al., 2015[14]) and earlier OECD projections (see Figure 8.A.1).

Figure 8.A.1. Global GHG emissions projections compared to the literature

Gt CO2-equivalent, including LULUCF

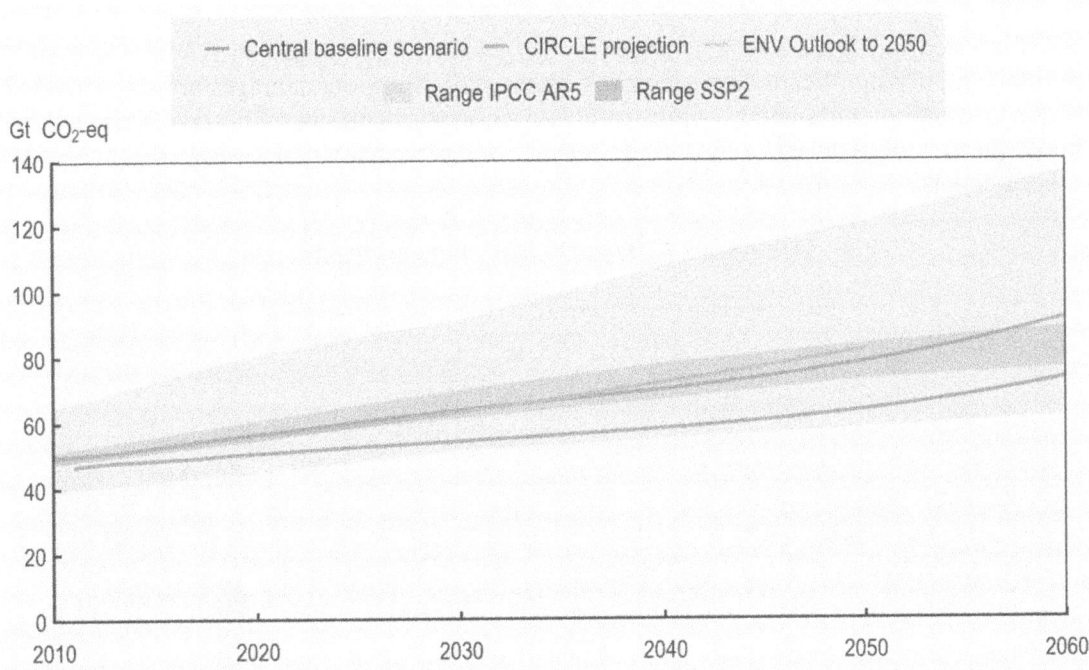

Note: Emissions from land use, land use change and forestry, as projected in (OECD, 2012[46]), added to the ENV-Linkages scenario to ensure comparability.
Source: Environment Outlook to 2050 (OECD, 2012[15]); CIRCLE projections (OECD, 2015[16]); Range IPCC Fifth Assessment Report (IPCC, 2014[13]); Range SSP2 (Riahi et al., 2015[14]).

The range of scenarios underlying the analysis of the IPCC is quite wide, and spans many different models and socioeconomic and technical assumptions, but exclude scenarios with (new) climate policies. The central baseline scenario falls well within this range, and is at the lower end of it until 2050. The projections are well aligned with the more recent SSP2 baseline projections, which represent socioeconomic trends that are similar to but not identical to the central baseline scenario. The trends for the coming decades are slightly lower than the range of the SSP2 projections, reflecting that economic growth projections for the short and medium term have been modified downward in recent years. In the longer run, the central baseline scenario assumes a relatively fast increase in emissions.

The central baseline scenario emissions projections are also slightly lower than earlier OECD projections, mostly because of lower projected fossil energy use. This is driven by recent developments, such as increased penetration of renewables through rapidly declining production costs (IEA, 2017[10]) and relatively low economic growth in recent years.

8.A.2. Climate damages

The economic damages from climate change that are induced by the emission projections of the central baseline scenario can be calculated using the methodology and data sources for the different climate damage endpoints of OECD (2015[16]). It should be acknowledged that this is only a partial assessment of the economic consequences of climate change, as many important impacts – including most extreme weather events, losses to biodiversity and ecosystems and large-scale singular events – could not be quantified.

As illustrated in Figure 8.A.2, damages are projected to be highest in Africa and Asia, with potentially significant downside risks. In the OECD region, damages are projected to be on average much smaller, while in a few cases (Canada and Russia), the positive consequences of reduced cold-related impacts are projected to outweigh the negative other impacts.

Figure 8.A.2. Damages from climate change

Percentage change with respect to central baseline scenario

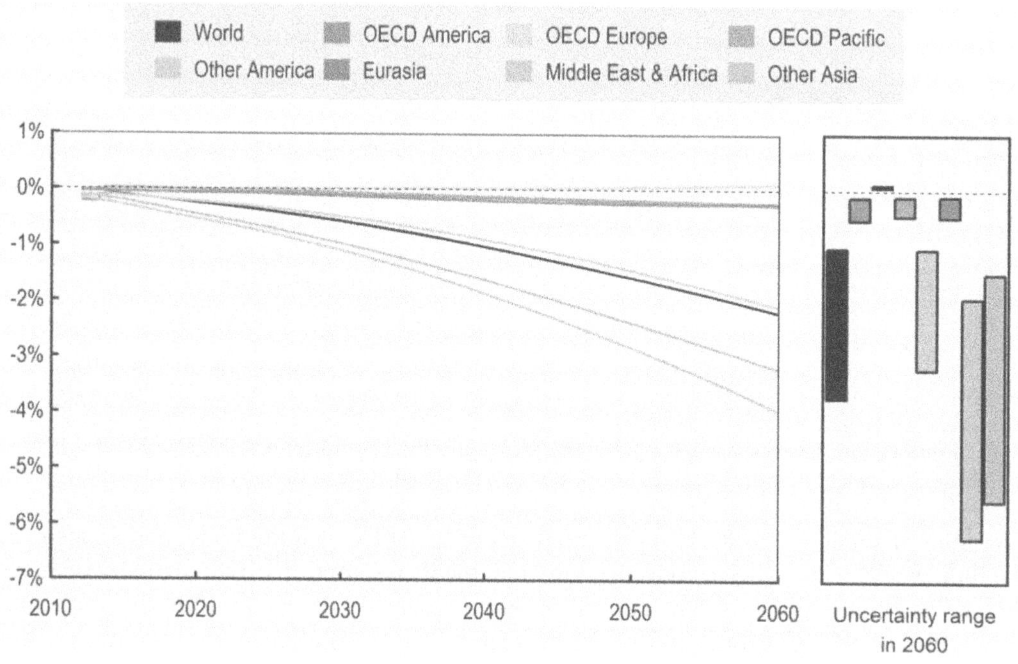

Note: OECD (2015[16]) describes the methodology to calculate these damages.
Source: OECD ENV-Linkages model.

At the global level, damages gradually increase to 1.1-3.7% of GDP by 2060; where the range reflects different levels of equilibrium climate sensitivity used for the calculation of the temperature changes from the emission concentrations.

The update of climate damages as presented in Figure 8.A.2 compare closely to the results discussed in detail in OECD (2015[16]). The specific numbers differ only relatively little from OECD (2015[16]) and mostly reflect the changes in the underlying economy baseline scenario. The methodology used is identical and the revision in the emissions projection is relatively minor. The lower emissions profile is compensated by a larger weight of emerging and developing economies in global GDP (due to the updated PPP exchange rates). Thus, the more vulnerable regions have a larger weight in global damages. This is reflected in a broader range of damage estimates at the world level: in the new central baseline scenario it amounts to 1.1-3.7% of GDP by 2060, whereas in OECD (2015[16]) this range was projected to be 1.0-3.3% of GDP.

8.A.3. Detailed explanations of the analysis of environmental impacts

What is Life Cycle Assessment?

Life Cycle Assessment (LCA) is a method generally used to assess environmental pressure related to a functional unit (product or service). The method has been developed since the early 1990s and has been standardised to a significant degree (ISO 14040 series, UNEP-SETAC Life Cycle Initiative (https://www.lifecycleinitiative.org/). Environmental impacts are assessed "from cradle-to-grave", from the point of extracting resources via production and use all the way down to waste disposal.

LCA consists of a number of steps. The steps wherein quantification takes place are the Life Cycle Inventory (LCI) and the Life Cycle Impact Assessment (LCIA).

In the LCI, process data is collected and combined. Such data can come out of standard LCI databases, but can also be collected specifically to serve the purpose of the LCA study. Processes are described in terms of their inputs (extracted resources and manufactured inputs coming out of another process) and outputs (emissions to the environment, waste, and the intended output of the process). The analysis uses the Ecoinvent database v2.2 (www.ecoinvent.org) together with the CMLCA software (www.cmlca.eu). The end result of the LCI is a comprehensive list of all extractions and emissions attributable to the selected functional unit. This list is often very long – it can contain hundreds or even thousands of items. This list is then the input for the LCIA.

The purpose of the LCIA is to translate the emissions and extractions into environmental impact. Different methodologies are used for LCIA. The UNEP-SETAC Life Cycle Initiative specifies a number of generally accepted approaches, one of which is used here: the method as specified by (Guinée, 2002[17]) in their Handbook to the ISO standard. This LCIA method translates emissions of substances into potential contributions to impact categories by using equivalency factors. This approach is well known from climate and greenhouse gas assessments: emissions of greenhouse gases (CO_2, CH_4, N_2O or others) are translated into CO2-equivalents based on their climate forcing properties. This enables to add them up into one impact category, in this case global warming. Emissions are translated by using equivalency factors into potential contributions to these impact categories. A list of equivalency factors used for aggregating emissions into different impact categories for roughly 2000 substances is available at www.universiteitleiden.nl/en/research/research-output/science/cml-ia-characterisation-factors.

Description of the life-cycle analysis impact categories and indicators

Environmental impacts are calculated for the extraction and processing of metals, concrete as well as sand and gravel. As described in Guinee (2002[17]), for each impact category an indicator is measured at the level of environmental pressure (interventions), i.e, emission of substance, extraction of resource and land use, because pressure indicators are directly linked to the process chain.

To be able to aggregate different pressure indicators (e.g. emissions of different GHG) into one impact category (climate change), the different pressure indicators are aggregated using so-called characterization factors as weights. The characterization factors express the relative contribution of an intervention to an impact category. The characterization factors are expressed relative to the impact of a reference emission or extraction. For example in the case of climate change the Global Warming Potential for a 100-year time horizon (GWP_{100}) is expressed relative to the 'infrared radiative forcing' of carbon dioxide and thus expressed as kg carbon dioxide equivalents per kg GHG emission.

Climate change is defined here as the impact of anthropogenic emissions on the radiative forcing of the atmosphere, causing the temperature of the earth's surface to rise. This leads to several impacts on the environment (e.g. sea level rise, extreme weather events, etc.) and the economy (e.g. agriculture and ecosystem services). The impacts are measured as emissions of GHG to the air (in kg CO_2-equivalents). These emissions are translated into a category indicator 'infrared radiative forcing' using the 100-year global warming potential (GWP_{100}) of different GHG (IPCC, 2013[18]).

Acidification is the corrosive impact that pollutants such as sulphur dioxide (SO_2) and Nitrous Oxides (NO_x) have on soil, groundwater, surface waters, biological organisms, ecosystems and materials (buildings). The impacts are measured as emissions of acidifying gases to the air (in kg SO_2 equivalents). These emissions are translated into an indicator 'deposition/acidification critical load', describing the fate and deposition of acidifying substances as Acidifying Potentials ($AP_{average\ Europe}$) of different gases (Huijbregts, 1999[19]).

Euthrophication covers all potential impacts of excessively high environmental levels of macronutrients, the most important of which are nitrogen (N) and phosphorus (P). Nutrient enrichment may cause undesirable shift in species composition and elevated biomass production in ecosystems and affects sources suitable for drinking water. The impacts are measured as emissions of nutrients to air, water and soil (in kg PO_4-equivalents). These emissions are translated into a category indicator 'deposition/N/P equivalents in biomass' using a stoichiometric procedure, which identifies the equivalence between N and P for both terrestrial and aquatic systems (Heijungs et al., 1992[20]).

Photochemical oxidation is the formation of reactive chemical compounds such as ozone by the action of sunlight on certain primary air pollutants, sometimes visible as smog. These reactive compounds may harm damage health, ecosystems, and crops. The impacts are measured as emissions of substances (VOC, CO) to air (in kg ethylene equivalents). These emissions are translated into a category indicator 'tropospheric ozone formation' using the Photochemical Ozone Creation Potential (POCP) of different gases (Jenkin and Hayman, 1999[21]; Derwent et al., 1998[22]; Derwent, Jenkin and Saunders, 1996[23]).

Human toxicity covers the impacts on human health of toxic substances in the environment, either by inhalation or via the food chain. Such impacts cover widely varying symptoms reaching from irritation to mortality. The impacts are measured as emissions of toxic substances to air, water and soil (in kg 1,4-dichlorobenzene equivalents). These emissions are translated into a category indicator 'acceptable daily intake/predicted daily intake' using Human Toxicity Potentials (HTP) (Huijbregts, 2000[24]; Huijbregts, 1999[25]).

Freshwater aquatic ecotoxicity refers to the impacts of toxic substances on species in freshwater aquatic ecosystems. The impacts are measured as emissions of toxic substances to air, water and soil (in kg 1,4-dichlorobenzene equivalents). These emissions are translated into a category indicator 'predicted environmental concentration/predicted no-effect concentration' using Freshwater Aquatic Ecotoxicity Potentials (FAETP) (Huijbregts, 2000[24]; Huijbregts, 1999[25]).

Terrestrial ecotoxicity refers to the impacts of toxic substances on species in terrestrial ecosystems. The impacts are measured as emissions of toxic substances to air, water and soil (in kg 1,4-dichlorobenzene equivalents). These emissions are translated into a category indicator 'predicted environmental concentration/predicted no-effect concentration' using the USES 2.0 model developed by RIVM, describing fate, exposure and effects of toxic substances into Terrestrial Ecotoxicity Potentials (TETP) (Huijbregts, 2000[24]; Huijbregts, 1999[25]).

Cumulative energy demand refers to the total energy use along the production chain of a material. It is also sometimes referred to as "energy footprint". The impacts are measured as energy use (in Joule). This energy use is summed into the category indicator 'cumulative energy demand' without any additional weighting of the different stages.

Land use refers the land surface used to produce the resource, for example the area occupied by a mine. This land is then temporarily unavailable for other uses, or for nature and ecosystems. The impacts are measured as land use (in m^2).

For some impact categories, a lot of information is available and a fair level of agreement is reached in circles of experts. Climate change, acidification, and eutrophication are among those. For other impact categories the uncertainties are larger. This can be due to limited data availability, but also to disagreement on the methodology to translate emissions or extractions into characterisation factors. Examples of highly uncertain impact categories are the different toxicity categories. To assess toxicity in LCA, fate models are commonly used in combination with no-effect levels, which are simplifications and in some cases lead to inadequate results. Emissions of metals in particular are uncertain. Toxicity is a highly complex impact category encompassing thousands of substances and many different mechanisms for impacting health or the environment.

The method used here to assess the environmental impacts of seven major metals is described by Van der Voet et al. (2018[11]). It differs from a "regular" LCA in a number of aspects.

In the first place, a "cradle-to-gate" approach is used, which specifies the impacts related to extraction and production, until the moment the material leaves the "gate" to be applied in all its different products. This is a commonly used approach when assessing materials. Impacts further in the life cycle are left out because it is no longer possible to attribute them to the individual materials that make up a product. Cradle-to-gate assessments therefore to not include the whole life cycle, only the part that is clearly related to the material itself. In that sense these assessments, although always an underestimation, provide relevant information for a more sustainable materials production system.

In the second place, the assumptions used for upscaling and forward-looking in time give this approach some characteristics of a more encompassing Life Cycle Sustainability Assessment (LCSA) approach.

The steps taken are the following. First, cradle-to-gate impacts are assessed for 1 kg metal for the present, using LCA as described above. Second, the evolution of the impacts per kg is assessed as a result of relevant developments, including the expected changes in supply routes for the different metals, in efficiency, in ore grades, in the energy background system. Third, the time specific impact factors are multiplied with the projected production to reach the projection of total environmental impacts.

8.A.4. Detailed results for projected environmental impacts

Table 8.A.1. Projected environmental impacts of selected metals in 2060

Indicator	Unit	Metal type	Aluminium	Copper	Iron	Zinc	Lead	Nickel	Manganese
Metal use*	Mt	Primary	139	57	2940	38	13	5	36
		Secondary	71	10	1133	4	18	2	16
Climate Change	Mt CO_2-eq	Primary	1771	436	6093	179	42	133	84
		Secondary	45	16	292	1	9	1	1
Acidification	kt SO_2 eq	Primary	5007	6589	23344	1933	861	776	449
		Secondary	229	158	1035	12	211	5	4
Eutrophication	kt PO_4- eq.	Primary	1275	54852	13059	1524	499	1648	126
		Secondary	145	121	521	5	19	1	2
Photochemical oxidation	kt ethylene eq.	Primary	572	166	3599	62	29	29	15
		Secondary	10	6	112	1	10	0	0
Human toxicity	Mt 1,4-dichlorobenzene eq.	Primary	7936	20278	25646	708	207	994	970
		Secondary	109	66	718	2	6	10	1
Freshwater aquatic ecotoxicity	Mt 1,4-dichlorobenzene eq.	Primary	900	13978	5793	339	115	621	102
		Secondary	35	32	945	1	6	2	0
Terrestrial ecotoxicity	Mt 1,4-dichlorobenzene eq.	Primary	4	5	36	7	0	2	0
		Secondary	1	0	64	0	0	0	0
Cumulative energy demand	PJ	Primary	24366	7309	89296	2605	497	1807	1413
		Secondary	789	237	5192	111	155	13	12
Land use	km²	Primary	26641	79575	232135	6891	2403	5117	5250
		Secondary	2828	832	16710	140	349	22	116

Note: * Metal use is expressed here in weight of refined materials use, not weight of raw ores to match the definitions of the per kg impacts.
Source: CMLCA model, based on the central baseline scenario of the OECD ENV-Linkages model.

CHAPTER 8. PROJECTIONS OF THE ENVIRONMENTAL CONSEQUENCES OF MATERIALS USE

Figure 8.A.3. Environmental impacts of selected metals are in most cases projected to more than double by 2060

Total environmental impacts in 2015 (lighter shaded area) and 2060 (full coloured area), index 1 for most polluting material in 2060

Figure 8.A.3. Environmental impacts of selected metals are in most cases projected to more than double by 2060 (*continued*)

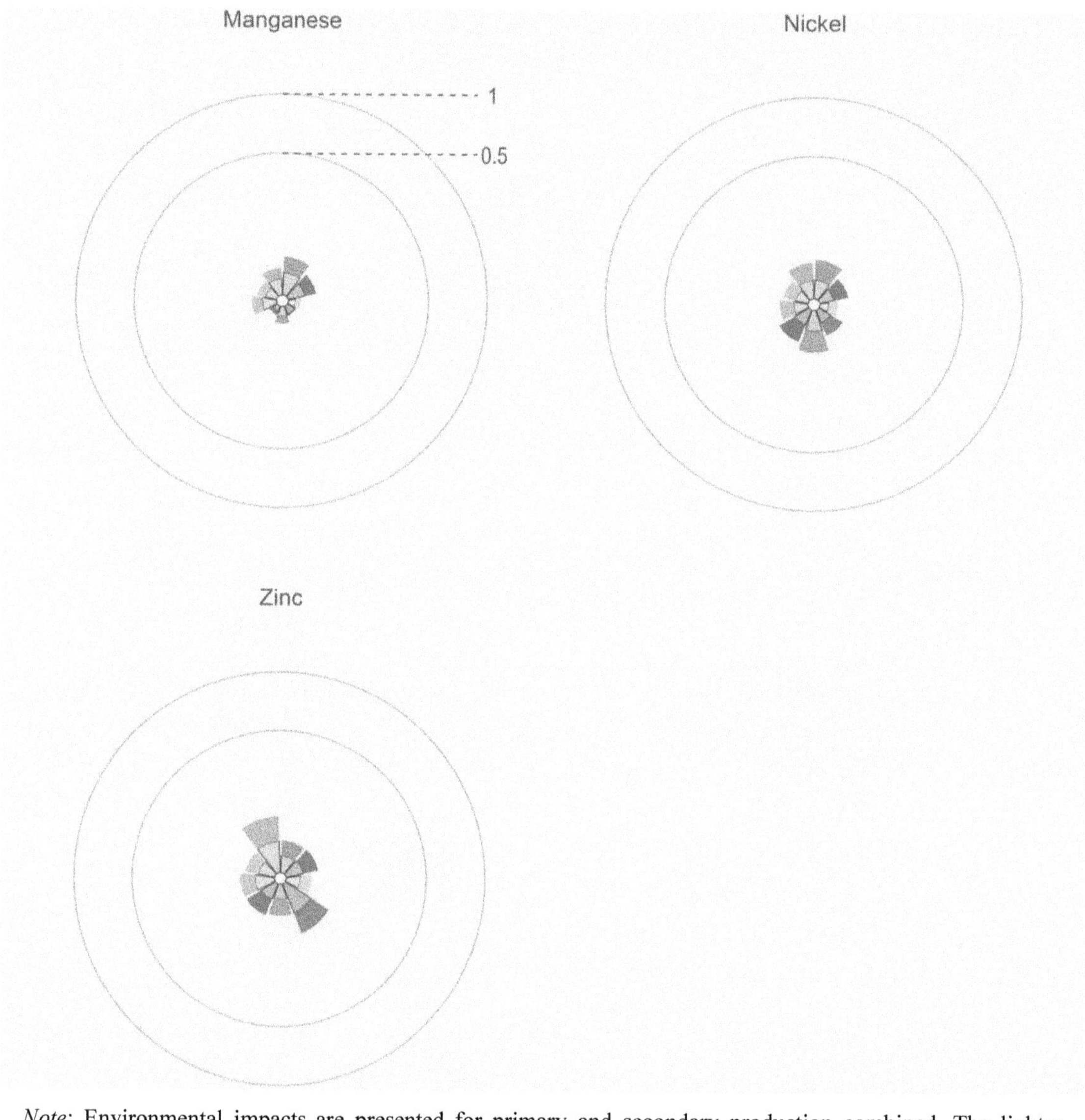

Note: Environmental impacts are presented for primary and secondary production combined. The lighter shading represents the value in 2015; the full coloured area reflects values in 2060. Impacts for "Other metals" reflect the combined impacts of lead, manganese, nickel and zinc.
Source: CML's CMLCA model, based on the central baseline scenario of the OECD ENV-Linkages model.

ORGANISATION FOR ECONOMIC CO-OPERATION AND DEVELOPMENT

The OECD is a unique forum where governments work together to address the economic, social and environmental challenges of globalisation. The OECD is also at the forefront of efforts to understand and to help governments respond to new developments and concerns, such as corporate governance, the information economy and the challenges of an ageing population. The Organisation provides a setting where governments can compare policy experiences, seek answers to common problems, identify good practice and work to co-ordinate domestic and international policies.

The OECD member countries are: Australia, Austria, Belgium, Canada, Chile, the Czech Republic, Denmark, Estonia, Finland, France, Germany, Greece, Hungary, Iceland, Ireland, Israel, Italy, Japan, Korea, Latvia, Lithuania, Luxembourg, Mexico, the Netherlands, New Zealand, Norway, Poland, Portugal, the Slovak Republic, Slovenia, Spain, Sweden, Switzerland, Turkey, the United Kingdom and the United States. The European Union takes part in the work of the OECD.

OECD Publishing disseminates widely the results of the Organisation's statistics gathering and research on economic, social and environmental issues, as well as the conventions, guidelines and standards agreed by its members.

www.ingramcontent.com/pod-product-compliance
Ingram Content Group UK Ltd.
Pitfield, Milton Keynes, MK11 3LW, UK
UKHW050413240426
12048UKWH00020B/1492

9 789264 307445